My STEM Workbook 2

Understanding
Science, Technology, Engineering and
Mathematics through design-process activities

Good Health & Well-being
Zero Hunger   Sustainable Cities and Communities
Clean Water and Sanitation   Climate Action
Affordable and Clean Energy
Life Below Water
Life on Land

This workbook belongs to

Years 3–4

Vinesh Chandra & Basil Slynko

My STEM Workbook 2
Understanding Science, Technology, Engineering and
Mathematics through design-process activities
Vinesh Chandra and Basil Slynko

Editor/Proofreader: Sandra Balonyi
Text designer: Michael Haddad
Cover designer: Michael Haddad; Concept: Basil Slynko
Illustrator: Michael Haddad

First published in Australia in 2023
Copyright ©2023 Vinesh Chandra and Basil Slynko
B. Slynko (Challenges 1, 5, 6, 8); V. Chandra (Challenges 2, 3, 4, 7)

National Library of Australia Cataloguing-in-Publication Data

Chandra, Vinesh and Slynko, Basil.
My STEM Workbook 2
Understanding Science, Technology, Engineering and Mathematics through design-process activities

ISBN: 978-0-6484052-3-8
For primary school age.

Printed in Australia
1 2 3 4 5 6 7 8 9 30 29 28 27 26 25 24 23

My STEM Workbook 2 – Understanding Science, Technology, Engineering and Mathematics through design-process activities Vinesh Chandra and Basil Slynko ISBN: 978-0-6484052-3-8

Contents

Challenge Activities

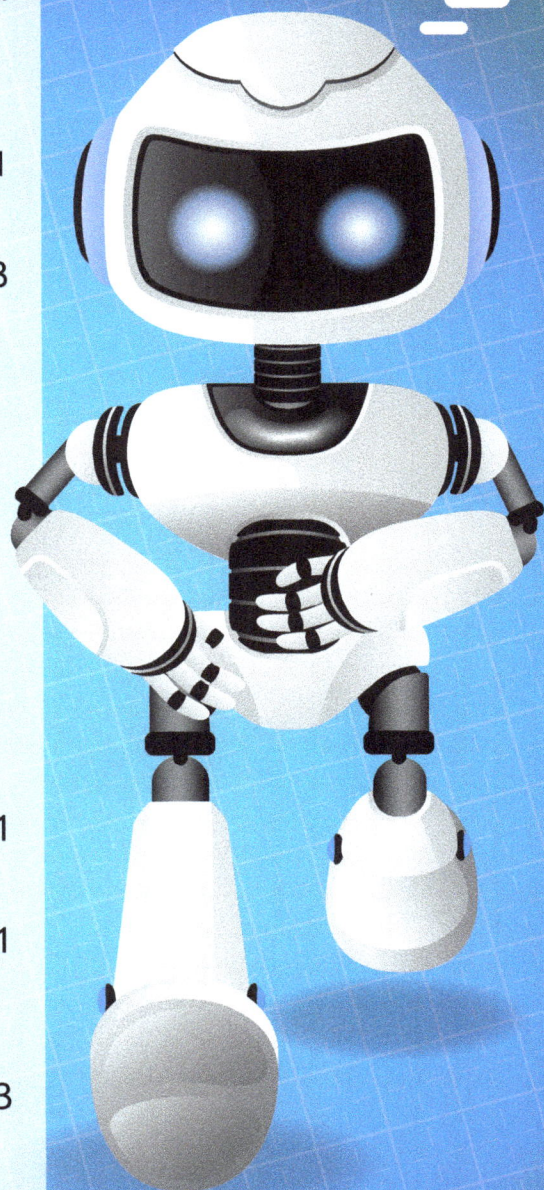

I ♥ STEM

Acknowledgements

The authors and publisher would like to credit or acknowledge the following sources for permission to use copyright material:

Adobe Stock: p1 (turmeric, ginger, lemongrass); p7 (PPE); p8 (herbs and spices); p15 (mask, handshake, coughing, contaminated surface); p15 (paper, cardboard, aluminium foil, leather, fabric, plastic film, leaf); p21 (scissors, needle, PPE); p33, p45, p55, p65 (PPE); p74 (underwater habitat); p82 (chicken).

Shutterstock: Robots (all pages); p21 (rotary knife); p26 (dry river bed); p40 (land yacht); p49 (urban sprawl); p51 (tiny homes, tiny home on wheels); p 66 (jar lights, tyre chairs, metal sculptures).

Illustrations p16 and p93: Paul Lennon

Every attempt has been made to trace and acknowledge copyright holders. Where the attempt has been unsuccessful, the publisher welcomes information that would redress the situation.

A special "thank you!" to Sandra Balonyi and Michael Haddad for working with us on this project.

Responsibility for errors remains with the authors.

About the authors

Associate Professor Vinesh Chandra is a university lecturer and teacher with more than 40 years of experience. He has taught in Australia and several countries overseas. His teaching areas include STEM, science, mathematics and technologies. His co-authored book titled *STEM Education in the Primary School: A Teacher's Toolkit* received two awards (Educational Publishing Australia Awards, 2021). Associate Professor Chandra's team won Gold in the prestigious QS Reimagine Education Wharton Awards in 2022 for their project in STEM Education in Papua New Guinea.

Basil Slynko, aka Professor Baz, B Ed St., MA; Design and Technologies educator – primary, secondary and tertiary – in Australia and overseas; Project-based-learning advocate; Curriculum consultant; Industry Experience – Construction and Manufacturing; Author and co-author of 25 titles, including *Nelson Introducing Technology Fourth Edition, Nelson Technology Activity Manual Third Edition,* and *Design and Technology in Today's World: A First Look.*

The United Nations aims to find ways to make our planet a better place for all.

Introduction

Humans have always relied on Science, Technology, Engineering and Mathematics, or STEM, to find solutions to challenges. Future generations will need to holistically draw upon this within and beyond their contexts. For example, STEM knowledge and skills are vital when addressing the United Nations Sustainable Development Goals.

What is the United Nations?
Many years ago, nearly all countries in the world joined to form an organisation called the United Nations.

What are the United Nations Sustainable Development Goals?
The United Nations Sustainable Development Goals are about improving the lives of people and all other living things to make our planet a better place. They are also about what we can do to care for our Earth so that future generations can also have a happy and healthy life.

My STEM Workbook 2 is part of a trilogy of STEM workbooks for primary students Years 1–6. Each workbook has eight challenges. Each addresses one of the United Nations Sustainable Development Goals (SDGs). Students apply their knowledge and skills of Science, Technologies, Engineering and Mathematics to propose solutions to contextually appropriate real-world challenges. These activities align with a range of content descriptors mandated in the Australian Curriculum Science, Technologies, and Mathematics (Version 9). However, these activities can also be implemented in other contexts, guided by other curriculum documents.

*Member states and territories of the United Nations

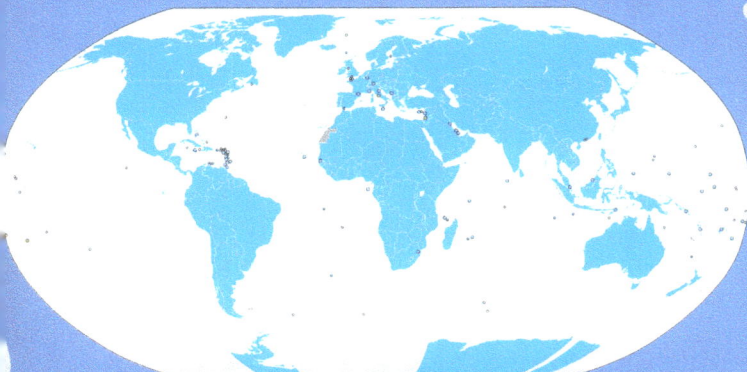

Notes for the teacher

The essence of integrated STEM education is project-based learning (PBL). It is a fun way for students to learn and teach. The real-world challenge in each activity is highly likely to interest, engage and enthuse students. *My STEM Workbook 2* comprises eight hands-on design activities, where students apply their STEM knowledge and skills to propose solutions to real-world challenges informed by the United Nations Sustainable Development Goals (SDGs). Each challenge is associated with an occupation. This should set the scene for inviting guest presenters, watching online vidoes and promoting class discussions.

The STEM PBL framework[1] was used to design the activities. Students need to tackle each challenge through the following steps:

Ask → Imagine → Plan → Create → Improve

Through these steps students apply their design-thinking skills to propose solutions to real-world challenges. Workplace Health and Safety is an integral part of each activity. Students are expected to handle tools, equipment and materials with care. Teachers are also expected to reinforce the use of safety gear – that is, Personal Protective Equipment (PPE) – as needed.

The trilogy of STEM Workbooks has a website (https://mystemworkbook.com/) which presents ideas on how knowledge from the digital technologies curriculum can be embedded within each challenge. The website also has support materials, including a video commentary on each activity.

1. Forbes, A., Chandra, V., Pfeiffer, L., & Sheffield, R. (2021). *STEM education in the primary school: a teacher's toolkit.* Cambridge University Press.

Connections to the Australian Curriculum
(Science, Design and Technologies, and Mathematics)

Subject	Content Descriptions	Activity							
		1	2	3	4	5	6	7	8
SCIENCE	compare characteristics of living and non-living things and examine the differences between the life cycles of plants and animals (AC9S3U01)								×
	compare the observable properties of soils, rocks and minerals and investigate why they are important Earth resources (AC9S3U02)	×							
	identify sources of heat energy and examine how temperature changes when heat energy is transferred from one object to another (AC9S3U03)			×		×			
	investigate the observable properties of solids and liquids and how adding or removing heat energy leads to a change of state (AC9S3U04)			×					
	consider how people use scientific explanations to meet a need or solve a problem (AC9S3H02)	×	×	×	×	×	×	×	×
	pose questions to explore observed simple patterns and relationships and make predictions based on experiences (AC9S1I01)	×	×	×	×	×	×	×	×
	use provided scaffolds to plan and conduct investigations to answer questions or test predictions, including identifying the elements of fair tests, and considering the safe use of materials and equipment (AC9S3I02)	×	×	×	×	×	×	×	×
	follow procedures to make and record observations, including making formal measurements using familiar scaled instruments and using digital tools as appropriate (AC9S3I03)	×	×	×	×	×	×	×	×
	construct and use representations, including tables, simple column graphs and visual or physical models, to organise data and information, show simple relationships and identify patterns (AC9S3I04)	×	×	×	×	×	×	×	×
	compare findings with those of others, consider if investigations were fair, identify questions for further investigation and draw conclusions (AC9S3I05)	×	×	×	×	×	×	×	×
	write and create texts to communicate findings and ideas for identified purposes and audiences, using scientific vocabulary and digital tools as appropriate (AC9S3I06)	×	×	×	×	×	×	×	×
	explain the roles and interactions of consumers, producers and decomposers within a habitat and how food chains represent feeding relationships (AC9S4U01)							×	
	identify sources of water and describe key processes in the water cycle, including movement of water through the sky, landscape and ocean; precipitation; evaporation; and condensation (AC9S4U02)			×					
	identify how forces can be exerted by one object on another and investigate the effect of frictional, gravitational and magnetic forces on the motion of objects (AC9S4U03)					×			
	examine the properties of natural and man-made materials including fibres, metals, glass and plastics and consider how these properties influence their use (AC9S4U04)		×			×	×		
	consider how people use scientific explanations to meet a need or solve a problem (AC9S4H02)	×	×	×	×	×	×	×	×
	pose questions to explore observed simple patterns and relationships and make predictions based on experiences (AC9S4I01)	×	×	×	×	×	×	×	×
	use provided scaffolds to plan and conduct investigations to answer questions or test predictions, including identifying the elements of fair tests, and considering the safe use of materials and equipment (AC9S4I02)	×	×	×	×	×	×	×	×
	follow procedures to make and record observations, including making formal measurements using familiar scaled instruments and using digital tools as appropriate (AC9S4I03)	×	×	×	×	×	×	×	×

Subject	Content Descriptions	Activity							
		1	2	3	4	5	6	7	8
DESIGN AND TECHNOLOGIES	examine design and technologies occupations and factors including sustainability that impact on the design of products, services and environments to meet community needs (AC9TDE4K01)	×	×	×	×	×	×	×	×
	describe how forces and the properties of materials affect function in a product or system (AC9TDE4K02)				×		×	×	×
	describe the ways of producing food and fibre (AC9TDE4K03)	×							×
	describe the ways food can be selected and prepared for healthy eating (AC9TDE4K04)	×							
	explore needs or opportunities for designing, and test materials, components, tools, equipment and processes needed to create designed solutions (AC9TDE4P01)	×	×	×	×	×	×	×	×
	generate and communicate design ideas and decisions using appropriate attributions, technical terms and graphical representation techniques, including using digital tools (AC9TDE4P02)	×	×	×	×	×	×	×	×
	select and use materials, components, tools, equipment and techniques to safely make designed solutions (AC9TDE4P03)	×	×	×	×	×	×	×	×
	use given or co-developed design criteria including sustainability to evaluate design ideas and solutions (AC9TDE4P04)	×	×	×	×	×	×	×	×
	sequence steps to individually and collaboratively make designed solutions (AC9TDE4P05)	×	×	×	×	×	×	×	×
MATHEMATICS	estimate the quantity of objects in collections and make estimates when solving problems to determine the reasonableness of calculations (AC9M3N05)	×	×	×	×	×	×	×	×
	use mathematical modelling to solve practical problems involving additive and multiplicative situations including financial contexts; formulate problems using number sentences and choose calculation strategies, using digital tools where appropriate; interpret and communicate solutions in terms of the situation (AC9M3N06)	×	×	×	×	×	×	×	×
	identify which metric units are used to measure everyday items; use measurements of familiar items and known units to make estimates (AC9M3M01)	×		×	×	×	×		×
	measure and compare objects using familiar metric units of length, mass and capacity, and instruments with labelled markings (AC9M3M02)	×		×	×	×	×		×
	recognise and use the relationship between formal units of time including days, hours, minutes and seconds to estimate and compare the duration of events (AC9M3M03)				×				
	identify angles as measures of turn and compare angles with right angles in everyday situations (AC9M3M05)	×	×	×	×	×			×
	recognise the relationships between dollars and cents and represent money values in different ways (AC9M3M06)								×
	make, compare and classify objects, identifying key features and explaining why these features make them suited to their uses (AC9M3SP01)	×	×	×	×	×	×	×	×
	interpret and create two-dimensional representations of familiar environments, locating key landmarks and objects relative to each other (AC9M3SP02)	×	×	×	×	×	×	×	×
	acquire data for categorical and discrete numerical variables to address a question of interest or purpose by observing, collecting and accessing data sets; record the data using appropriate methods including frequency tables and spreadsheets (AC9M3ST01)				×	×			
	create and compare different graphical representations of data sets including using software where appropriate; interpret the data in terms of the context (AC9M3ST02)				×	×			
	conduct guided statistical investigations involving the collection, representation and interpretation of data for categorical and discrete numerical variables with respect to questions of interest (AC9M3ST03)							×	
	choose and use estimation and rounding to check and explain the reasonableness of calculations including the results of financial transactions (AC9M4N07)	×	×	×	×	×	×		×
	use mathematical modelling to solve practical problems involving additive and multiplicative situations including financial contexts; formulate the problems using number sentences and choose efficient calculation strategies, using digital tools where appropriate; interpret and communicate solutions in terms of the situation (AC9M4N08)	×	×	×	×	×	×		×
	interpret unmarked and partial units when measuring and comparing attributes of length, mass, capacity, duration and temperature, using scaled and digital instruments and appropriate units (AC9M4M01)	×	×	×	×	×	×		×
	recognise ways of measuring and approximating the perimeter and area of shapes and enclosed spaces, using appropriate formal and informal units (AC9M4M02)						×		
	estimate and compare angles using angle names including acute, obtuse, straight angle, reflex and revolution, and recognise their relationship to a right angle (AC9M4M04)	×	×		×	×	×		×
	represent and approximate composite shapes and objects in the environment, using combinations of familiar shapes and objects (AC9M4SP01)	×	×		×	×	×		×
	acquire data for categorical and discrete numerical variables to address a question of interest or purpose using digital tools; represent data using many-to-one pictographs, column graphs and other displays or visualisations; interpret and discuss the information that has been created (AC9M4ST01)				×		×		
	analyse the effectiveness of different displays or visualisations in illustrating and comparing data distributions, then discuss the shape of distributions and the variation in the data (AC9M4ST02)				×			×	
	conduct statistical investigations, collecting data through survey responses and other methods; record and display data using digital tools; interpret the data and communicate the results (AC9M4ST03)						×	×	

Challenge Activities

Notes for the student

In this workbook, you will engage in eight activities. In each activity, you will use science, technology, engineering and mathematics, or STEM, to develop solutions to real-world challenges.

What will I do in the Challenge activities?

In each challenge activity you will follow these steps:

Step 1. Ask
What is the challenge?

Step 2. Imagine
How can I tackle the challenge?

Step 3. Plan
How can I plan my idea?

Step 4. Create
What will my idea look like?

Step 5. Improve
How can I make my idea better?

Then, at the end, you will **reflect** on the activity and its outcomes to see what you have learnt.

The use of digital technologies is encouraged. Ask your teacher.

Pages 95–98 at the back of this workbook provide some additional practice grids for you to test your ideas at the planning stage and space for notes. Use the blank page if you wish to record any interesting things as you go along.

Now, let's have a look at the eight Challenge activities. Each activity is based on a different United Nations Sustainable Development Goal…

Challenge 1
– A herb and spice community garden
Goal: Zero Hunger

Challenge 2
– A mask to fight infectious diseases
Goal: Good Health & Well-being

Challenge 3
– A water-cycle experiment
Goal: Clean Water and Sanitation

Challenge 4
– A yacht that moves on land
Goal: Affordable and Clean Energy

Challenge 5
– A tiny house
Goal: Sustainable Cities and Communities

Challenge 6
– An item made from waste
Goal: Climate Action

Challenge 7
– A digital book on plastics in waterways
Goal: Life below Water

Challenge 8
– A shelter for chickens
Goal: Life on Land

Challenge 1

A herb and spice community garden

United Nations Sustainable Development Goal:

Zero Hunger

Zero **hunger** is one of the goals of the United Nations. The zero-hunger goal is about finding ways to make sure that everyone has enough to eat. One of the ways to do this is by growing your own food. Herbs and spices are one example of food that can be grown. Herbs and spices add flavour to meals prepared around the world.

Chefs in your local community need an area to grow herbs and spices. Your class has been asked to plan and make a garden. Your challenge is to find a suitable location for growing herbs and spices in the school grounds. The **garden** could be any size – a plot of land or a collection of portable trays.

1. Ask

Here are three examples of herbs and spices.

Turmeric   *Ginger*   *Lemongrass*

Q1a. What herbs and spices could/does the community use? Write your answers here.

_____

_____

_____

_____

_____

_____

Key words in this activity:

- garden
- hunger
- ground
- minerals
- bird's-eye view
- plant
- shape
- soil

Q1b. What is one similarity and one difference between herbs and spices? Use the Word Bank to complete these sentences.

Similarity:

Both herbs and spices are _____ .

Difference:

Herbs are usually the _____
of a **plant**.

Spices are generally the _____, _____

and _____ *of a plant.*

Q1c. Give an example of each.

An example of a herb is _____

An example of a spice is _____

Q1d. List the three main components of **soil**.

1. _____

2. _____

3. _____

Q1e. Why do plants need **soil**?

_____

_____

_____

_____

_____

Q1f. Why do plants need **minerals**?

_____

_____

_____

Word Bank

flowers   fruit   leaves
plants   seeds

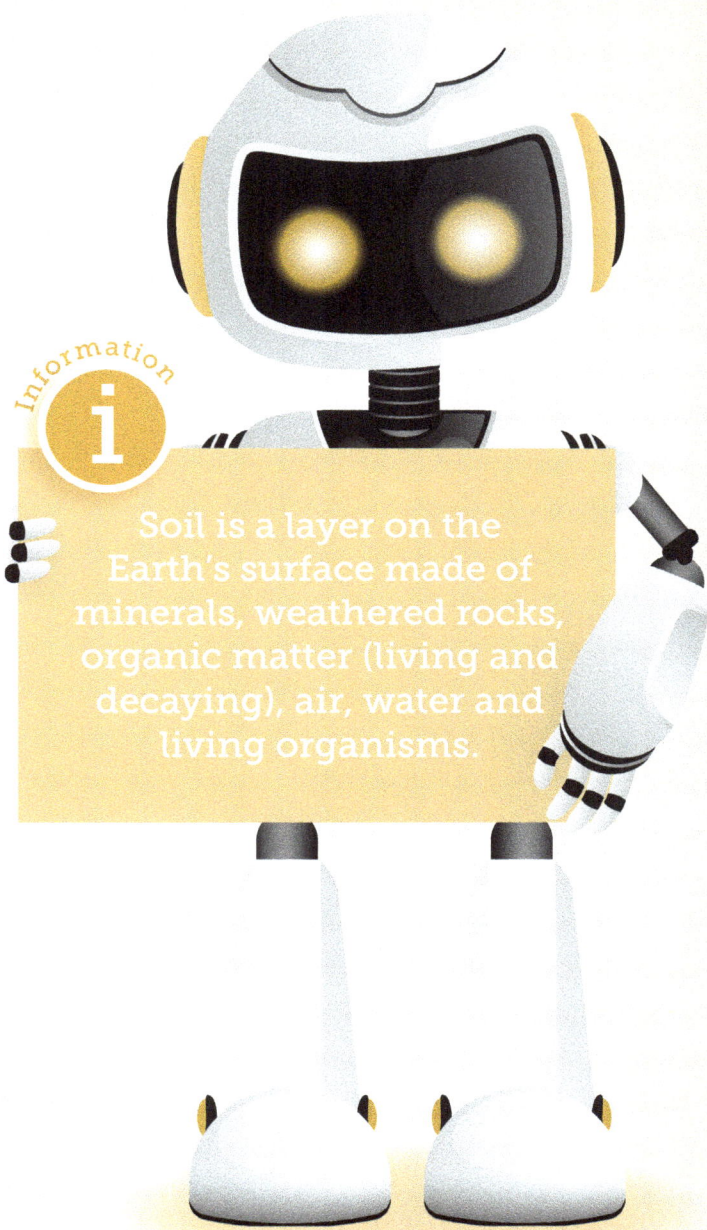

Information

Soil is a layer on the Earth's surface made of minerals, weathered rocks, organic matter (living and decaying), air, water and living organisms.

2

r Read about herbs and spices and how to grow them.

Research

You can gather information from books in your library or you can do a search on the Internet. Here are some websites and search terms to get you started.

1. YouTube: https://www.youtube.com/watch?v=ZpfQHI6i6X0
   Video Title: Herbs and Spices for Kids in English to Learn
2. PBS: https://www.pbs.org/parents/crafts-and-experiments/grow-an-herb-garden
   Search terms: Grow an herb garden PBS
3. SEE Library: https://seelibrary.org/ Look up the "Science and Technology" category.
   Search terms: seelibrary, science and technology

i *Information*

Herbs and spices need certain things to grow. Some of the basic needs are sunlight, water, space and soil.

Q1g. Some people use potting mix to grow herbs and spices. Why?

_____
_____
_____
_____
_____
_____
_____

Q1h. Answer the following questions about the growing needs of herbs and spices.

• How much sunlight do they need to grow?

_____
_____

• How much water do they need to grow?

_____
_____

• How much space do they need to grow?

_____
_____

• What type of soil do they need to grow?

Herbs need _____

Spices need _____

• What type of fertiliser(s) do they need?

Herbs need _____

Spices need _____

2. Imagine

Herbs and spices need the right weather conditions. For example, mint grows best in warmer climates.

Q2a. What herbs and spices can be grown in your local community garden/area given the weather conditions? Use the Internet to research what herbs and spices can be grown. Share your ideas with your teacher.

_____

_____

_____

_____

_____

_____

_____

Q2b. List three herbs/spices that you would plant. What do they need to grow?

Herb/spice name	What it needs to grow
1	
2	
3	

Information

Sometimes soil needs to be improved before herbs and spices can grow in it.

Q2c. How can soil be improved so that herbs and spices can grow in it?

_____

_____

_____

_____

_____

_____

_____

_____

_____

_____

_____

_____

_____

_____

Challenge 1. A herb and spice community garden

3. Plan

Q3a. Use the grid to draw your initial **bird's-eye view** of your herb and spice garden. Show the layout of the herbs and spices.

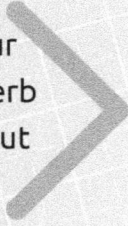

Q3b. Share your herb and spice garden idea with your classmates. Some things to consider are:
- **shape**
- size (in metres or centimetres)
- access to water
- safety of public users
- green waste
- fertiliser.

Ask them for their comments and write them here.

_____

_____

_____

_____

_____

_____

_____

_____

_____

_____

_____

→ *Consider the feedback...*

Q3c. After considering your classmates' comments draw your final bird's-eye view of your garden and the layout of the herbs and spices. Remember to:
- label the materials you will need
- label the herbs and spices you want to grow
- include the measurements of the garden and the layout of the herbs and spices
- add any other information and details.

*The **shape** of my garden is:*

_____

Q3d. Use this table to list the herbs and spices and how many (quantity) you will need as well as any other materials and equipment you will need for your herb and spice garden.

Herbs and spices	Quantity

Materials and equipment

→ Procedure

Q3e. What procedure will you use to make your garden? Draw a line from each process to the correct step number.

Add mulch to the garden

Mark out the planting

Get the herbs and spices

Get the soil

Mark out the layout

Add the soil

Plant the herbs and spices

Select a site

Step

1
2
3
4
5
6
7
8

Challenge 1. A herb and spice community garden

Tools & Equipment

te

Gardening tools are used to work the herb and spice garden. Some gardening tools are shown below.

Secateurs

Shovel

Rake

Saw

Garden fork

Hoe

Safety

S

Safety is important to prevent accidents and injuries.
Using gardening tools and equipment can be dangerous if they are used improperly.
You should always wear safety gear to protect yourself when gardening.
The other term for safety gear is "Personal Protective Equipment" or "PPE".

Some examples of PPE: safety glasses, gardening gloves and gumboots.

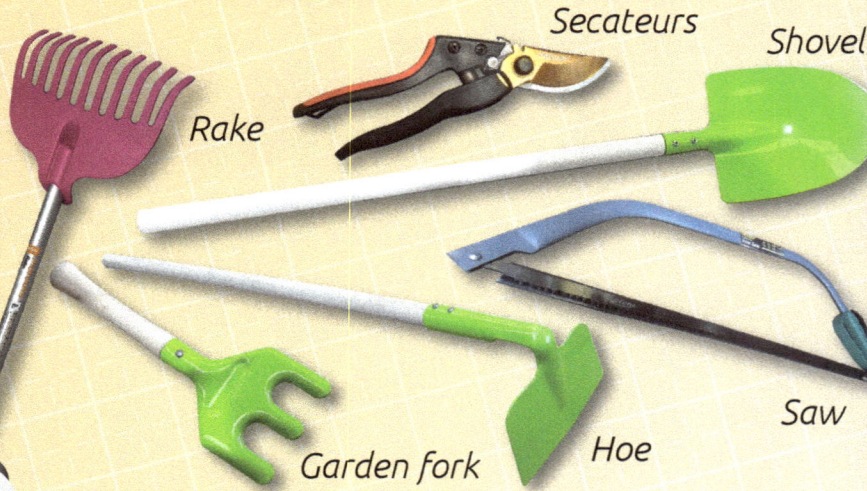

Q3f. Select three gardening tools from those shown above and write your selection below. Then, write how each tool could injure you and what safety gear you should wear to prevent the injury.

1 Gardening tool:_____

How can I get injured?_____

_____

What safety gear should I wear?_____

_____

2 Gardening tool:_____

How can I get injured?_____

_____

What safety gear should I wear?_____

_____

3 Gardening tool:_____

How can I get injured?_____

_____

What safety gear should I wear?_____

_____

4. Create

Making a herb and spice garden

Remember to ask questions and seek feedback from your teacher while you are creating your herb and spice garden.

You need to:

- gather the herbs and spices for your community garden
- gather the tools and equipment you need
- follow your procedure steps to make the herb and spice garden
- use PPE and follow safety rules.

Q4a. Making your herb and spice garden may not go as planned. For each step that caused problems, explain how you dealt with it. List the step(s) and your course of action.

_____
_____
_____
_____
_____
_____
_____
_____
_____
_____
_____
_____
_____
_____
_____
_____

Fun Activity

Name these herbs and spices... Fill in the missing letters in the list below to spell out the phrase reading downwards.

t___yme
nutm___g
sor___el
___asil
par___ley
cori___nder
ci___namon
car___amom
ro___emary
pep___er
ch___lli
ma___e
cay___nne
lemongra___s
s___ge
___ill
mustar___
___enugreek
nige___la
tarrag___n
cher___il
saffr___n
c___min
ca___away
min___
marj___ram
tur___eric
ging___r
oreg___no
c___ove
chive___

8

Challenge 1. A herb and spice community garden

Observe

Watch your herb and spice garden grow...

Q4b. Observe the plants in your garden regularly. Use the space on these two pages and/or a separate journal to record your observations. Make a note of any changes that you notice in the plants. You can support your observations with sketches and/or digital images.

Observations:

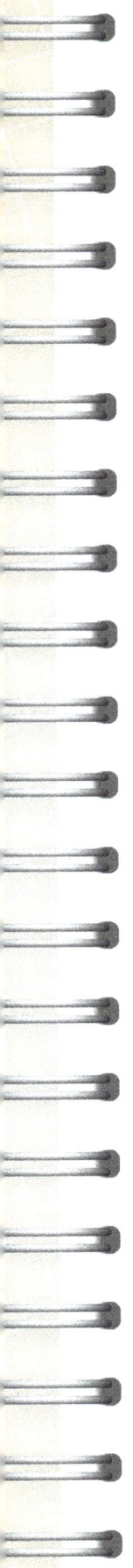

5. Improve

It is time to present your final herb and spice garden idea to your class. Explain how your idea supports the United Nations Goal of Zero Hunger.

Ask your classmates these questions:

- *What are some of the good points about my herb and spice garden idea?*
- *What are some of the weak points about my herb and spice garden idea?*

What did your classmates think?

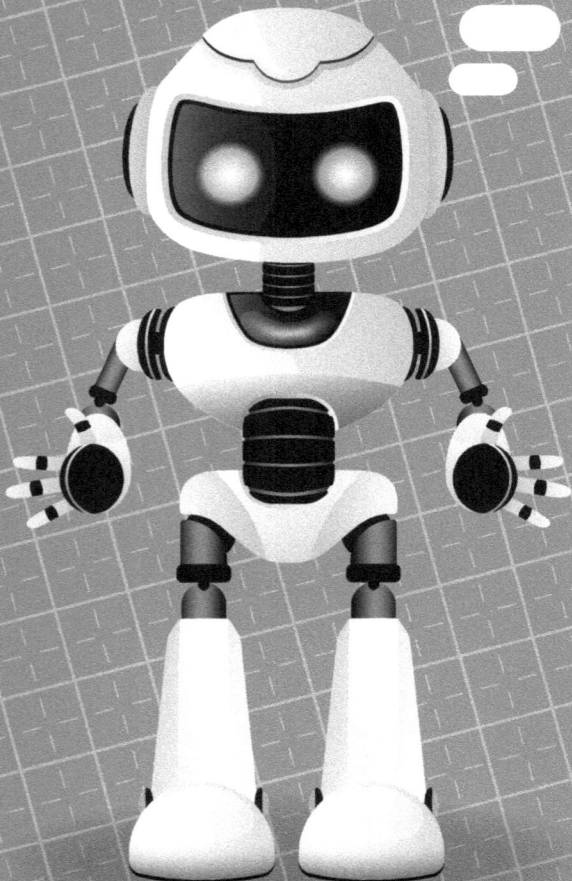

Q5a. Write your classmates' answers in the table.

Good points

Weak points

→ *Consider your classmates' feedback...*

Q5b. What could you do to improve your herb and spice garden idea?

_____

_____

_____

_____

_____

_____

_____

_____

Now it's time to Reflect...

What have you learnt in this activity? *List three things.*
Hint: Look back at the key words.

1. _____

2. _____

3. _____

What would you like to know more about?

_____
_____
_____
_____
_____
_____

An entrepreneur is someone who starts a business to raise the standard of living and create jobs.

Information

Curly Question

If your garden produces bumper crops, how could you sell the herbs and spices? What steps would you take to establish an entrepreneurial approach to selling bumper crops?

_____
_____
_____
_____
_____
_____
_____
_____
_____

Challenge 1. A herb and spice community garden

Challenge 2

A mask to fight infectious diseases

United Nations Sustainable Development Goal:

Good Health & Well-being

Coronavirus disease (COVID-19) is one example of an **infectious disease** that can be passed from one person to another. Millions of people around the world die each year because of **infectious diseases**. Finding ways to stop the spread of infectious diseases can help achieve the United Nations goal of promoting good health and well-being.

Key words in this activity:

• infectious disease
• microbes
• symmetrical
• prototype
• design

Your school nurse has asked your class to create prototypes of masks that can be used to prevent the spread of infectious diseases. Your **prototype** needs to be made from one piece of material without any joins. At the parents' open day, you will present your prototype and explain how the mask can prevent the spread of germs. *Remember: your mask is only a prototype. It cannot replace a real mask.*

1. Ask

Q1a. Sketch two examples of masks. What are their shapes?

The shape is

_____

The shape is

_____

r *search*

Read about masks and how they are made.
You can gather information from books in your library
or you can do a search on the Internet. Here are some
websites and search terms to get you started.

1. YouTube: https://www.youtube.com/watch?v=cZZ4AIQESLA
   Video Title: Masks for kids
2. SEE Library: https://seelibrary.org/ and look up the "Health" category.
   Search terms: seelibrary, science and technology

i *Information*

A mask needs to be symmetrical.

Q1b. What does **symmetrical** mean?
Write your answer below. Draw lines on the
sketches you made in Question 1a to show what
you mean.

_____

_____

_____

_____

Q1c. COVID-19 is an **infectious disease**. What does
this mean?

_____

_____

_____

_____

Q1d. How do masks prevent the spread of infectious diseases?

_____

_____

_____

2. Imagine

Q2a. Use the Word Bank to complete the following paragraph.

Germs are very tiny _____.

They are so tiny that we can only see them under a

_____. Sometimes germs are

also referred to as _____. People

get an _____ disease when

microbes *such as the coronavirus enter their bodies.*

Word Bank

infectious   microbes
microscope   organisms

> **Information**
>
> The transfer of microbes is also known as 'transmission'. There are three common ways that microbes can be transferred: direct contact, airborne, and contaminated surfaces and objects.

Direct contact: handshake

Airborne: coughing

Contaminated surfaces: touching

Q2b. This is an example of a mask used during COVID-19. List three **design** features of this mask.

1._____

2._____

3._____

Q2c. Consider these statements about masks. Are they correct or incorrect? Circle your answer.

i) *Masks can stop the spread of germs through direct contact.*
Correct | Incorrect

ii) *Masks can stop the spread of germs that are airborne.*
Correct | Incorrect

iii) *Masks can stop the spread of germs found on contaminated surfaces.*
Correct | Incorrect

Q2d. Masks can be made from a variety of materials. The table below lists some common properties. *State the importance of each property and give reasons.* You may want to look at some materials, such as a piece of cardboard or cloth, to understand more about these properties. Use the Internet and/or see page 93 *"Some properties of materials"*.

Property	Is the property important? Yes/No/Maybe	Reasons
strength		
toughness		
hardness		
durability		
absorbency		
biodegradability		
stiffness		

Q2e.
Each example below shows a property of materials. Can you label the property? (See page 93 for the answers.)

_____

_____

_____

_____

16

Q2f. Which of the materials listed below will be most suitable for making a mask? Give reasons for your answers.

Material	Is this material suitable? Yes/No	Reasons
 paper		
 cardboard		
 aluminium foil		
 leather		
 fabric		
 plastic film		
 large leaves		

17

Q2g. Rex believes fabric is the best material to use to make a mask. Do you agree? ◯ Yes ◯ No ◯ Maybe

Explain your answer: _____

_____

_____

_____

Q2h. List two fabrics that can be used to make a mask. Why are these fabrics good for making a mask?

*Fabric 1:*_____

This fabric is good for making a mask because _____

_____

_____

_____

*Fabric 2:*_____

This fabric is good for making a mask because _____

_____

_____

_____

Q2i. There are two basic groups of fabric types: natural and man-made. *List two examples of natural fabrics and two examples of man-made fabrics.*

Natural

1 _____

2 _____

Man-made

1 _____

2 _____

3. Plan

Q3a. Use the grid to draw a rough sketch of your initial mask idea.

>

Q3b. Have a good look at your sketch of a mask. Can your initial idea be improved on?
Some things to consider are:
- shape
- dimensions
- type of fabric
- comfort
- size.

Write your ideas for improvements here.

→ *Consider your improvements...*

Q3c. Draw a final design of your mask.
Remember to:
- label the parts of your mask
- label the dimensions of your mask (in centimetres)
- add any other information and details such as fabric type

Challenge 2. A mask to fight infectious diseases

Q3d. List the three key materials you will need to make your mask. How much will you need? Why will you use this material?

	Material	Size/Quantity	I will use this material because...
1			
2			
3			

→ **Procedure**

Q3e. What procedure will you use to make your mask? List the steps.
Remember to say how you will make your mask symmetrical.

_____	_____
_____	_____
_____	_____
_____	_____
_____	_____

Tools & Equipment

Three tools that can be used when working with fabric are scissors, a needle, and a rotary knife.

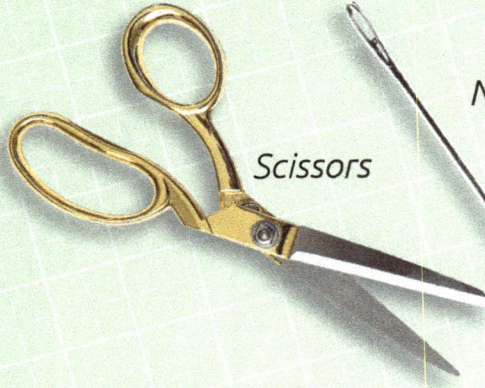
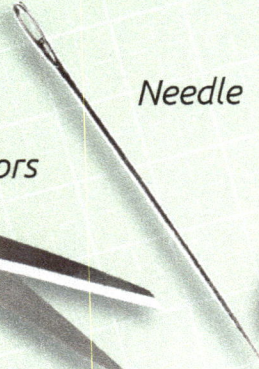
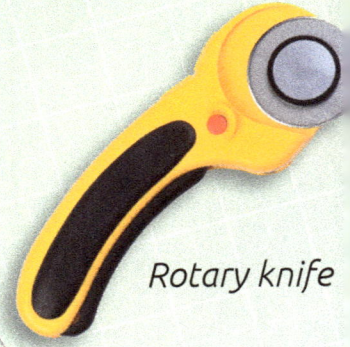

Scissors

Needle

Rotary knife

Safety

S

Safety is important to prevent accidents and injuries. Using tools and equipment can be dangerous. You should always wear safety gear to protect yourself. The other term for safety gear is "Personal Protective Equipment" or "PPE".

Some examples of PPE: an apron, ear muffs and a thimble.

Q3f. Look at the three tools above. Write how each tool can injure you and how you can also use it safely.

1 *How can I get injured with scissors?*_____

_____

_____

*How can I use scissors safely?*_____

_____

_____

2 *How can I get injured with a needle?* _____

_____

_____

*How can I use a needle safely?*_____

_____

_____

3 *How can I get injured with a rotary knife?*_____

_____

_____

*How can I use a rotary knife safely?*_____

_____

_____

4. Create
Making a mask

Remember to ask questions and seek feedback from your teacher while you create your mask.
You need to:

- gather the materials
- gather the tools and equipment you need
- follow your procedure steps to make the mask
- use PPE and follow safety rules.

Q4a. Making your mask may not go as planned. For each step that caused problems, explain how you dealt with it. List the step(s) and your course of action.

_____

_____

_____

_____

_____

_____

_____

_____

_____

_____

_____

_____

_____

Test

Testing is a very important part of any design process...

Q4b. How will you test your mask?

_____

_____

_____

_____

_____

_____

_____

_____

_____

_____

Q4c. Slogans and logos can help raise awareness of important issues. *Write a slogan or draw a graphic on this t-shirt* aimed at the importance of stopping the spread of infectious diseases. Use colour if it helps with your message.

22

Challenge 2. A mask to fight infectious diseases

5. Improve

It is time to present your mask to the class. Explain how your idea supports the United Nations Goal of Good Health & Well-being.

Ask your classmates these questions:

- *What are some of the good points about my mask idea?*
- *What are some of the weak points about my mask idea?*

What did your classmates think?

Q5a. Write your classmates' answers in the table.

Good points

Weak points

→ *Consider your classmates' feedback...*

Q5b. What could you do to improve your mask idea?

_____

_____

_____

_____

_____

_____

_____

_____

Now it's time to Reflect...

What have you learnt in this activity? *List three things.*
Hint: Look back at the key words.

1. _____

2. _____

3. _____

What would you like to know more about?

_____

_____

_____

_____

_____

_____

Curly Question

Can masks be made using smart materials? Give some examples.

_____

_____

_____

_____

_____

_____

_____

_____

24

Challenge 3

A water-cycle experiment

United Nations Sustainable Development Goal:

Clean Water and Sanitation

One of the goals of the United Nations is to ensure that everyone has access to clean water. We rely on the **water cycle** for clean water. Without the water cycle the amount of clean water on our planet will quickly decline.

A weather presenter wants you to plan and conduct an experiment to show how the water cycle works. You will gather and graph data to support your reasoning. In a class presentation, you will use your experiment set-up, **data** and **graph** to explain the water cycle.

1. Ask

Q1a. Why do we need clean water?

_____

_____

_____

_____

_____

Q1b. What does the term 'water cycle' mean to you?

_____

_____

_____

_____

_____

Key words in this activity:

• water cycle
• condensation
• evaporation
• data
• graph
• Celsius

Some rivers, lakes and dams dry up.

A dry river bed

Q1c. If a river, dam or lake dries up, where does the water go and what happens to it?

_____

_____

_____

_____

r **Read about how the water cycle works.**
You can gather information from books in your library or you can do a search on the Internet. Here are some websites and search terms to get you started.

1. Science Sparks: www.science-sparks.com/make-a-mini-water-cycle
   Search terms: Make a mini water cycle model
2. National Geographic Kids: www.natgeokids.com/au/discover/science/nature/water-cycle
   Search terms: The water cycle National Geographic kids
3. SEE Library: https://seelibrary.org/ and look up the "Science and Technology" category.
   Search terms: seelibrary, science and technology

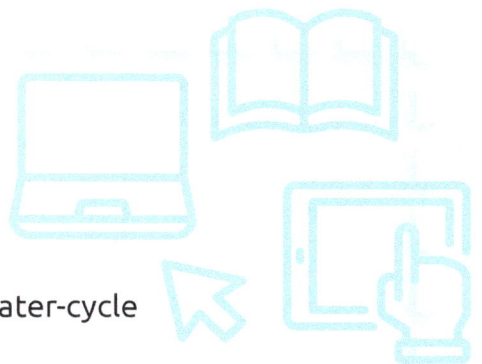

2. Imagine

Q2a. This water-cycle diagram shows how water moves around the Earth. Use the Word Bank to complete the boxes in the diagram.

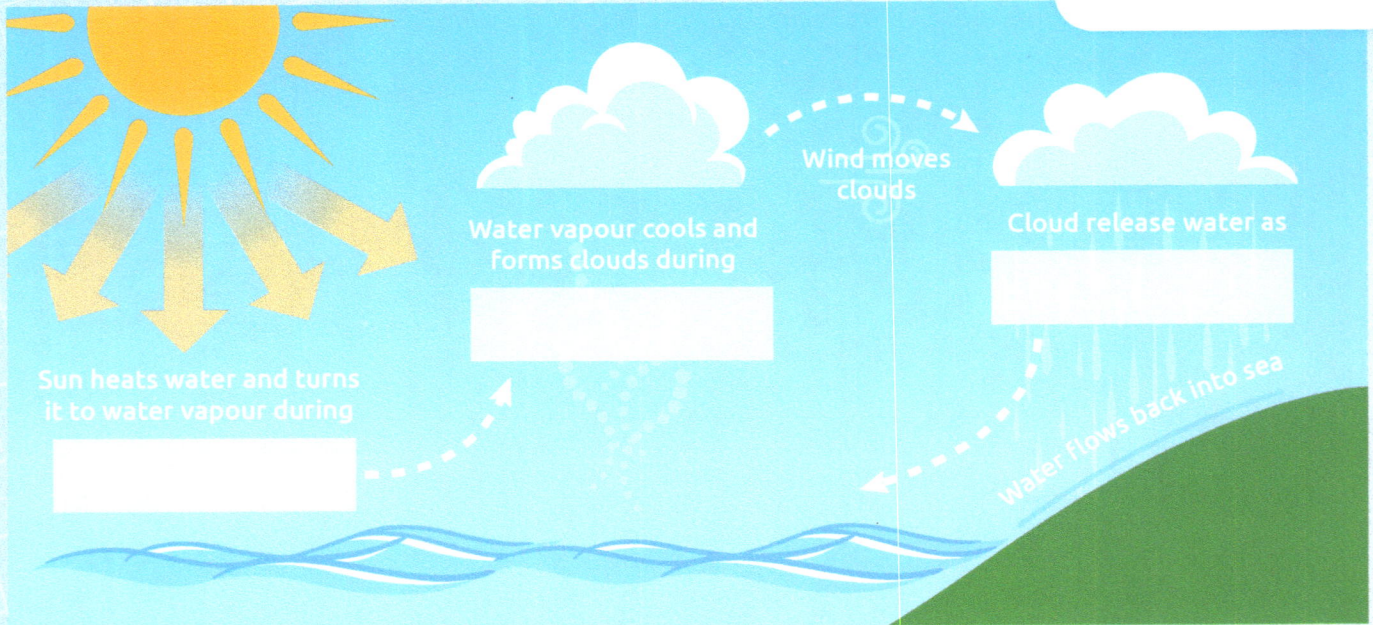

Wind moves clouds

Water vapour cools and forms clouds during

Cloud release water as

Sun heats water and turns it to water vapour during

Water flows back into sea

Q2b. Fill in the blanks.

Evaporation occurs when _____ heats up and changes to water _____.

Condensation occurs when water _____ cools down and forms _____.

Evaporation and condensation are two _____ processes.

Q2c. What is another name for precipitation? _____

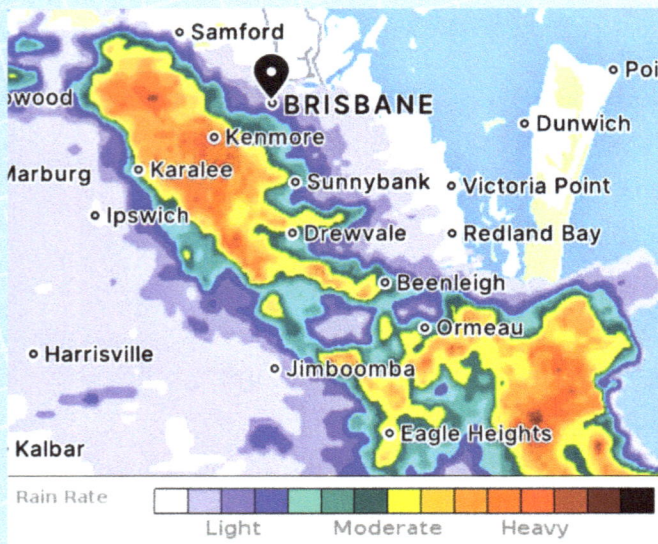

The Australian Bureau of Meteorology's radar images show the location of rain. The colours depict the intensity of rainfall.

Q2d. Using the radar map here can you identify an area which will experience:

Heavy rainfall _____

Moderate rainfall _____

Light rainfall _____

No rainfall _____

Rain Rate

Light   Moderate   Heavy

Source: Bureau of Meteorology. http//www.bom.gov.au

27

Challenge 3. A water-cycle experiment

Information

An experiment is a scientific procedure carried out to determine if a prediction is correct, and what the overall results are.

Jane set up an experiment to learn about evaporation and condensation. Here is a part of Jane's experimental report:

My water-cycle report

by Jane

Investigation question: What happens when water evaporates?

Prediction: Water condenses after evaporation.

Materials/equipment used: Two glass food containers of different sizes, a glass tumbler, plastic wrap, food colouring, thermometer.

Procedure: I did this experiment on a bright, sunny day. In the larger container, I added 50 millilitres of water and a drop of food colouring. I placed the empty small container inside the larger container. I then tightly covered the large container with some plastic wrap. I left my experiment on the lawn for a few hours. I also filled a glass with some water and placed a thermometer in it. I recorded the temperature of the water every 30 minutes. I measured the temperature in degrees Celsius.

Results: After a few hours, I observed that the small container had some clear water. The temperature of the water in the glass had also gone up.

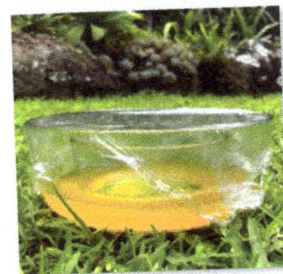

28

→ *Answer the following questions about Jane's experimental report...*

Q2e. Why did Jane do her experiment on a sunny day?

_____

_____

_____

Q2f. What was the smaller container used for in Jane's experiment?

_____

_____

_____

Q2g. Why did Jane add food colouring to the water?

_____

_____

_____

Q2h. After a few hours, Jane observed that there was some clear water in the smaller container. There were also some water droplets underneath the plastic wrap. How did this happen?

_____

_____

_____

_____

Q2i. If Jane had not covered the glass containers with the plastic wrap, what would have happened?

_____

_____

_____

Q2j. Why did Jane measure the temperature of the water in the glass?

_____

_____

_____

Q2k. What would be a good conclusion for Jane's experiment?

_____

_____

_____

Q2l. How will you set up your experiment to show the water cycle?

_____

_____

_____

_____

_____

_____

You might like to do some research to get some ideas.

3. Plan

Q3a. Use the grid to draw a sketch of how you intend to set up your water-cycle experiment.

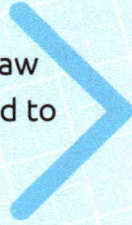

In your drawing, you need to:

• label the equipment you will use to set up the experiment
• label the materials you will need to gather data so it can be graphed
• add any other information and details.

Q3b. List the five key materials you will need for your experiment. What quantities will you need? Why will you use these materials?

Material	Quantity	I will use this material because...

→ *Procedure*

Q3c. Write the steps you will follow to conduct your experiment. Include steps that will enable you to measure temperature changes.

_____

_____

_____

_____

_____

_____

_____

_____

_____

_____

Safety

S Safety is important to prevent accidents and injuries. Using tools and equipment can be dangerous. You should always wear safety gear to protect yourself when conducting experiments. The other term for safety gear is "Personal Protective Equipment" or "PPE".

Some examples of PPE: protective gloves, face masks, ear plugs, apron, hard hat, protective glasses and gumboots.

Tools & Equipment

te

Q3d. In the table below, list any three items of equipment you will need to conduct your water-cycle experiment. Write how you can injure yourself and also how you can use the equipment safely.

Equipment	How I can injure myself	How I can be safe

4. Create

Conducting a water-cycle experiment

Remember to ask questions and seek feedback from your teacher while you are conducting your experiment.

You need to:

- gather the materials and the equipment you need
- follow your procedure to conduct the experiment
- remember to record the temperature
  (**Suggestion:** *record the temperature every 10 minutes for one hour.*)
- use PPE and follow safety rules.

Q4a. Record the time and temperature in this table.

Time (minutes)	0	10	20	30	40	50	60
Temperature (°C)							

Q4b. Using the data you have gathered complete the graph below. Provide a title and label the axes with units. Record what the graph is showing you.

Title _____

The graph shows:

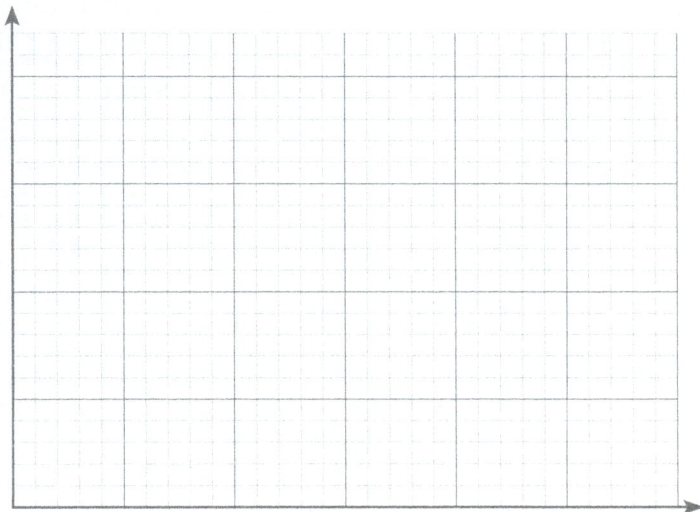

_____

_____

_____

_____

_____

_____

_____

_____

_____

Label (X axis): _____

_____

Label (Y axis): _____

_____

Q4c. Conducting your experiment may not go as planned. For each step that caused problems, explain how you dealt with it. List the step(s) and your course of action.

_____

_____

_____

_____

_____

_____

_____

_____

_____

5. Improve

Present your experiment and data to your class and explain how your idea relates to the United Nations Goal of Clean Water and Sanitation.

Ask your classmates these questions:

- *What are some of the good points about my experiment on how the water cycle works?*
- *What are some of the weak points about my experiment on how the water cycle works?*

What did your classmates think?

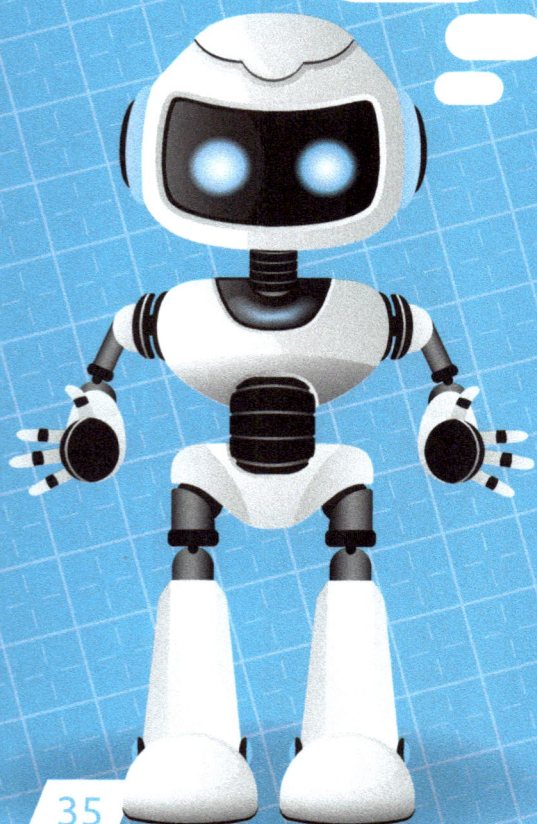

Q5a. Write your classmates' answers in the table.

Good points

Weak points

→ *Consider your classmates' feedback...*

Q5b. What could you do to improve your experiment and explanation?

_____

_____

_____

_____

_____

_____

_____

_____

Now it's time to Reflect...

What have you learnt in this activity? *List three things.*

Hint: Look back at the key words.

1. _____

2. _____

3. _____

What would you like to know more about?

_____

_____

_____

_____

_____

_____

Curly Question

What is acid rain? _____

_____

_____

_____

_____

What causes acid rain? _____

_____

_____

_____

_____

Challenge 4

A yacht that moves on land

United Nations Sustainable Development Goal:

Affordable and Clean Energy

One of the goals of the United Nations is to enable people all over the world to have access to affordable, clean and sustainable forms of energy. One of the ways of doing this is by using sustainable alternative energy sources that are free, such as the sun, wind and tides.

Key words in this activity:

- energy
- force
- friction
- land yacht
- design feature
- centimetres
- engineered system
- specifications

A mechanical engineer wants you to plan and build a model of a land yacht using mostly recyclable materials. Land yachts are powered by wind. Your land yacht should be no longer than 40cm and no wider than 20cm. You and your classmates will test your land yachts on different surfaces to understand the effect of **friction**. A pedestal fan will power your land yacht.

1. Ask

Q1a. What is a **land yacht**?

_____

_____

_____

_____

_____

_____

_____

_____

_____

Q1b. List two similarities and two differences between a yacht that sails on water and one that travels on land.

Similarities
1
2

Differences
1
2

Q1c. What is one feature that enables a land yacht to travel on land? Why is this feature needed?

_____

_____

_____

_____

_____

_____

_____

_____

_____

_____

_____

_____

_____

_____

_____

r **Read about how land yachts are powered.** You can gather information from books in your library or you can do a search on the Internet. Here are some websites and search terms to get you started.

1. YouTube: https://www.youtube.com/watch?v=qN0V0NXV3Kw
   Search terms: What is friction?
2. YouTube: https://www.youtube.com/watch?v=LSevw1sfpk
   Search terms: Why is it more difficult to pull a boat on the beach than on the sea
3. SEE Library: https://seelibrary.org/ Look up the "Science and Technology" category.
   Search terms: seelibrary, science and technology

It keeps getting stuck!

A **force** is a push or a pull. **Energy** is the ability to do work. **Work** is done when a force is applied to move an object.

2. Imagine

Phil can push the cart without any problems when it has wheels on. However, he cannot push the cart when the wheels are taken off. He gives up and says *"I cannot push this cart, it keeps getting stuck"*.

Q2a. Why can't Phil push the cart when the wheels are taken off?

_____

_____

_____

_____

_____

_____

Q2b. Apart from putting the wheels back on the cart, what else can Phil do to solve his problem?

_____

_____

_____

_____

_____

_____

Q2c. How does a land yacht work? Use the words 'force' and 'energy' in your answer.

_____

_____

_____

_____

_____

_____

_____

Information

A land yacht has a number of different systems. Each system has a number of parts which make it work.
Read more about Systems on page 42.

Q2d. Study this photograph. Identify each part from the Word Bank; then complete the table below.

Word Bank

frame   mast   sail
seat   steering   wheel

Part	Name of part	This part is needed because...
1		
2		
3		
4		
5		
6		

My STEM Workbook 2 – Understanding Science, Technology, Engineering and Mathematics through design-process activities Vinesh Chandra and Basil Slynko ISBN: 978-0-6484052-3-8

Q2e. List three **design features** of land yachts and why they are important.

*1 Design feature:*_____

   This feature is important because _____

   _____

*2 Design feature:*_____

   This feature is important because _____

   _____

*3 Design feature:*_____

   This feature is important because _____

   _____

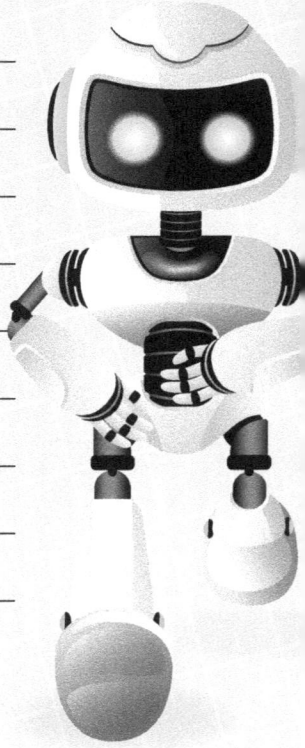

Q2f. What will the design **specifications** of your land yacht be?

The name of my land yacht will be _____.

The length will be _____ *centimetres (cm). The width will be* _____ *centimetres (cm).*

Recyclable materials I could use are _____

_____

I would use these materials because _____

_____

The body of my land yacht will be made of _____

because _____

For the wheels I will use _____

because _____

For the sail I will use _____

because _____

The sail will be attached to the mast using _____

For the axles I will use _____

because _____

I will/may need help with _____

_____

Information

More information about systems...

Do you know what a system is?

*A **system** is a set of parts, things or ideas that work together. A system is another way to create an outcome or complete a task.*

These are the basic parts of a system:

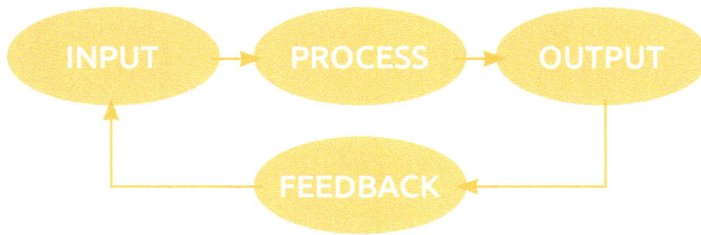

INPUT → PROCESS → OUTPUT

FEEDBACK

- The **input** is any resource that contributes to a system, such as information, people, materials and tools.
- The **process** is all the activities done to produce the result.
- The **output** is the result or goal and its impacts. The impacts may be benefits or costs.
- The **feedback** is the information – positive or negative – about the output. More resources may be required to improve the output.

Our world is full of systems. Examples are communication systems, transportation systems and manufacturing systems.

Systems are usually made up of a number of **sub-systems**. Each sub-system has its own input, process, output and feedback. Sub-systems may also use lower sub-systems that work together to produce the output.

Examples of sub-systems in motor vehicles:

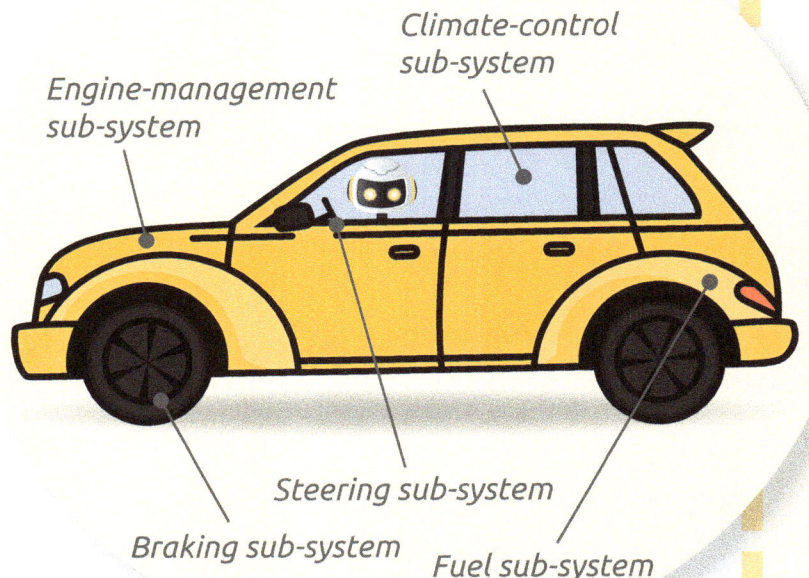

Climate-control sub-system

Engine-management sub-system

Steering sub-system

Braking sub-system

Fuel sub-system

Source: *Design and Technology in Today's World: A First Look, Years 3–6*; Basil Slynko.

Challenge 4. A yacht that moves on land

My STEM Workbook 2 – Understanding Science, Technology, Engineering and Mathematics through design-process activities   Vinesh Chandra and Basil Slynko   ISBN: 978-0-6484052-3-8

3. Plan

Q3a. Use the grid to draw a rough sketch of the land yacht you intend to create.

Q3b. Share your land-yacht idea with your classmates. Some things to consider are:
• shape
• dimension (in **centimetres**)
• type of material.
Write your ideas for improvements here.

→ *Consider your improvements...*

Q3c. Sketch a revised side view of your land yacht. Remember to:
• label the materials and parts
• label the dimensions of your land yacht (in centimetres)
• add any other information and details
• label the common shapes in your sketch.

Challenge 4. A yacht that moves on land

Q3d. List the four key materials you will need to build your land yacht. What quantities will you need? Why will you use this material?

Material	Quantity	I will use this material because...

→ Procedure

Q3e. What procedure will you follow to make your land yacht?
Remember you also need to gather data. List the steps.

_____   _____

_____   _____

_____   _____

_____   _____

_____   _____

_____   _____

_____   _____

_____   _____

4. Create

Making a model of a land yacht

Remember to ask questions and seek feedback from your teacher while you create your model.

You need to:

- gather the materials
- gather the tools and equipment you need
- follow your procedure to make the land yacht model
- use PPE and follow safety rules.

Q4a. Making your land yacht may not go as planned. For each step that caused problems, explain how you dealt with it. List the step(s) and your course of action.

_____

_____

_____

_____

_____

_____

_____

_____

_____

_____

_____

_____

_____

_____

_____

_____

_____

_____

S Safety is important to prevent accidents and injuries. Using tools and equipment can be dangerous. You should always wear safety gear to protect yourself. The other term for safety gear is "Personal Protective Equipment" or "PPE".

Some examples of PPE: face mask, earmuffs and safety glasses.

t *Test*

Q4b. Now test your yacht on three different surfaces and enter your results in the table.

Your teacher will show you the different surfaces where you will test your model. Your teacher will also set up a pedestal fan and show you how to measure the time it takes your yacht to reach the finish line.

The surfaces can be concrete, timber, grass, tiles and so on. In order to work out the best surface, you will need to note the time it takes from the start to the finish line. You may need a stopwatch.

Surface	Time

Q4c. Complete the following sentences.

The force of friction is greatest on

_____

because _____

_____

_____

_____ .

The force of friction is least on

_____

because _____

_____

_____

_____ .

fa *Fun Activity*

Unscramble these words... then copy the letters that fall on a number into the space below to spell out a key phrase. Can you also find them in the grid...?

sepcsor   _ _ _ 5 _ _ 21 _ _

rlevtienara   _ 15 _ _ _ _ 12 1 _ _ _ _ _

ogla   _ _ _ 9 _

ygeern   _ 20 _ _ 23 _

tibesuaslna   _ _ _ _ _ 7 _ _ 17 8 _ _

nidw   _ _ 18 6

dfkaebce   3 _ _ 13 _ _ 14 _

cfreo   2 _ _ _ _

okrw   _ _ 22 _

yrlcabecla   _ _ _ _ 24 _ _ 11 _ _ 16

tsesmy   _ _ _ _ _ 10

ropwe   _ _ _ 19 _

utoptu   4 _ _ _ _ _

```
P A Q K L R O V W J C G T P L S
R E C Y C L A B L E Z X A O U Y
O U D B H K L S L Q R Y Z W O S
C O R G L V T B C D P I F E T T
E P J S R A E D S V R F O R C E
S Y X E N E R G Y A G E W C Z M
S P V H N I N A E W O E E P D T
Z G O F G K A V B I S D R A M U
F O D S U S T A I N A B L E O P
R A G H W N I Q C D J A M Z N T
G L M O O S V U I S D C K B L U
O D V W I M E O W O R K L X Z O
```

_ _ _ _ _ _ _ _ _ _ _ _ _ _ _ _ _ _ _ _ _ _ _ _
1 2 3 4 5 6 7 8 9 10 11 12 13 14 15 16 17 18 19 20 21 22 23 24

5. Improve

What did your classmates think of your land yacht? Explain how your idea relates to the United Nations Goal of Affordable and Clean Energy.

Ask your classmates these questions:

- *What are some of the good points about my land yacht?*
- *What are some of the weak points about my land yacht?*

> What did your classmates think?

Q5a. Write your classmates' answers in the table.

Good points

Weak points

→ *Consider your classmates' feedback...*

Q5b. How could you improve your land yacht?

_____
_____
_____
_____
_____
_____
_____
_____

Now it's time to Reflect...

What have you learnt in this activity? *List three things.*
Hint: Look back at the key words.

1. _____

2. _____

3. _____

What would you like to know more about?

_____
_____
_____
_____
_____
_____

Curly Question

Should we allow wind-powered vehicles on our roads?
Explain your answer.

_____
_____
_____
_____
_____
_____
_____
_____

Challenge 5

A tiny house

United Nations Sustainable Development Goal:

Sustainable Cities and Communities

One of the goals of the United Nations is to take good care of our cities and communities. We can all have a better life if we look after our cities and communities. Urban sprawl is an issue as our cities grow. Land is being swallowed by buildings, especially houses. Reducing urban sprawl is a good thing to do.

An urban planner has challenged you to design and build a model of a tiny house. One way to deal with urban sprawl is to reduce the size of houses in cities. A growing trend in today's world is the rise of "tiny houses". Tiny houses are a solution to the ever-increasing costs of building a home and the sustainability of materials. The size of your model should be 70cm long, 30m wide and 25cm high per level – maximum 3 levels. This will represent 7m x 3m x 2.5m in the real world. Tiny-house designs and models are to be featured at an upcoming Sustainable Cities and Communities display in the City Mall.

[Key words in this activity:

- urban sprawl • area
- structure • estimate
- measurements
- insulation • glazing
- shading]

Urban sprawl

1. Ask

Q1a. What does a tiny house mean to you?

_____

_____

_____

_____

Q1b. What do you need to consider when designing a tiny house? *Circle the three things that you think are **most** important from the list below.*

size   shape   materials

use   windows   doors

insulation   position   weather

> *Information*
>
> **i**
>
> **Materials can be sorted into two groups: natural and man-made.**
>
> *See page 94, "Some samples of natural and man-made materials".*

Q1c. What natural materials could you use to make your tiny house? *List them here.*

_____   _____

_____   _____

_____   _____

_____   _____

Q1d. What man-made materials could you use to make your tiny house? *List them here.*

_____   _____

_____   _____

_____   _____

_____   _____

Research

r

Read about tiny houses and the materials used to construct them.

You can gather information from books in your library or you can do a search on the Internet. Here are some websites and search terms to get you started.

1. NASA: https://climatekids.nasa.gov/tiny-houses/
   Search terms: Tiny houses for kids
2. PBL Works: https://www.pblworks.org/video-tiny-house-project
   Search terms: Tiny Houses Project PBL
3. SEE Library: https://seelibrary.org/ Look up the "Science and Technology" category.
   Search terms: seelibrary, science and technology

2. Imagine

Tiny House Design

Building a tiny house is about the use of space. Space is not wasted. Walkways are wide enough to use. Stairs are compact and often spiral in design. Space below the stairs is often used for storage and a place for other things such as a toilet where there is enough height. Living **areas**, kitchen and bathrooms that require plumbing are grouped. They could be on one level or placed above each other.

An open-plan layout is common. This layout allows the space to be used for a number of purposes – for example, a kitchen, dining and living areas. The kitchen can be isolated by using slide-out screens, creating a wall, and hiding the appliances and cabinets when not in use.

Interior walls are kept to a minimum. Interior walls are often made of opaque materials. That way, light can enter the room and privacy is achieved. The outside walls may use opaque or translucent materials to enable light to penetrate the building. Translucent materials enable the occupants to see life outside the home. Screens or slats on the exterior of translucent materials aid privacy.

The **structure** may be narrow, but the distinct levels, including the roof, add living spaces to a tiny house. A vertical void through various levels, roof to ground, allows the light to flood the interior spaces of each level and create a feeling of the outdoors.

All materials have a set of properties such as strength, hardness, or toughness These properties determine how a material should be used. *See page 93 "Some properties of materials".*

Tiny houses in the Netherlands

A tiny house on wheels

Q2a. List three interesting features of these houses that you could use in your tiny-house design.

1. _____

2. _____

3. _____

Q2b. Tiny houses can be built using natural and man-made materials. These are listed in the table. *Record the advantages and disadvantages of each material. Are they natural or man-made?*

Material	Natural or man-made	Advantages	Disadvantages
Wood for the frame			
Sheet metal for the roof			
Plywood for the floor			
Timber sheeting for interior walls			

Q2c. What materials could you use to create the various parts of your model? *Complete the table below.*

Material for a tiny house	Suitable material to make my tiny-house model
Frame e.g. wood	
Roof e.g. sheet metal	
Floor e.g. plywood	

Information

Plywood is made from thin sheets of wood glued together. Sheets are at right angles to each other.

52

3. Plan

Q3a. Use the grid to draw a rough sketch of your initial tiny-house idea.

Q3b. Share your tiny-house idea with your classmates. Some things to consider are:
• materials
• size and shape
• layout of living areas
• access
• visual appeal.
Ask them for their comments and write them here.

_____

_____

_____

_____

_____

_____

_____

_____

_____

_____

_____

_____

→ *Consider your classmates' feedback...*

Q3c. After considering your classmates' comments draw a final front view sketch of your tiny-house model. Remember to:
• label the materials you will need
• include the dimensions (length, width, height)
• add any other information and details.

Q3d. Use this table to list the materials and quantities (No.) you will need for your tiny-house model idea.

Material (N or MM)*	No.	Material (N or MM)*	No.

* N = Natural material   MM = Man-made material

→ **Procedure**

Q3e. How will you make your tiny-house model?
Write down the steps you will follow.

Safety

S Safety is important to prevent accidents and injuries. Using tools and equipment can be dangerous when making a model of your tiny house. You should always wear safety gear to protect yourself. The other term for safety gear is "Personal Protective Equipment" or "PPE".

Some examples of PPE: safety glasses, disposable gloves and an apron.

Q3f. A few construction processes used to build a tiny house are listed in the table below. *List the health hazard(s) of each process and what PPE you would wear to prevent each health hazard.*

Process	Health hazard(s)		PPE I should wear
Cutting sheet metal			
Cutting plywood			
Fixing timber sheeting			
Preparing walls for painting			

4. Create
Making a model of a tiny house

Remember to ask questions and seek feedback from your teacher while you create your model.
You need to:

- gather the materials: paper, cardboard and paddle-pop/ice-cream sticks
- gather the tools: scissors, stapler, and paper/bulldog clips
- gather the equipment and adhesives: glue and adhesive tape
- follow your procedure steps to make the tiny-house model
- use colour to render your model
- use PPE and follow safety rules.

Q4a. Making your tiny-house model may not go as planned. For each step that caused problems, explain how you dealt with it. List the step(s) and your course of action.

_____

_____

_____

_____

_____

_____

_____

_____

_____

_____

_____

Q4b. Calculate the perimeters and areas of your tiny-house model.
Use the spaces provided below.

Perimeter of the floor
Area of the floor
Perimeter of the front wall
Area of the front wall
Perimeter of the side wall
Area of the side wall
Perimeter of the roof
Area of the roof

Information

i

Sunlight heats the ground and things on the Earth. As the position of the sun changes, the temperatures of the seasons also vary, and so do the size of the shadows.

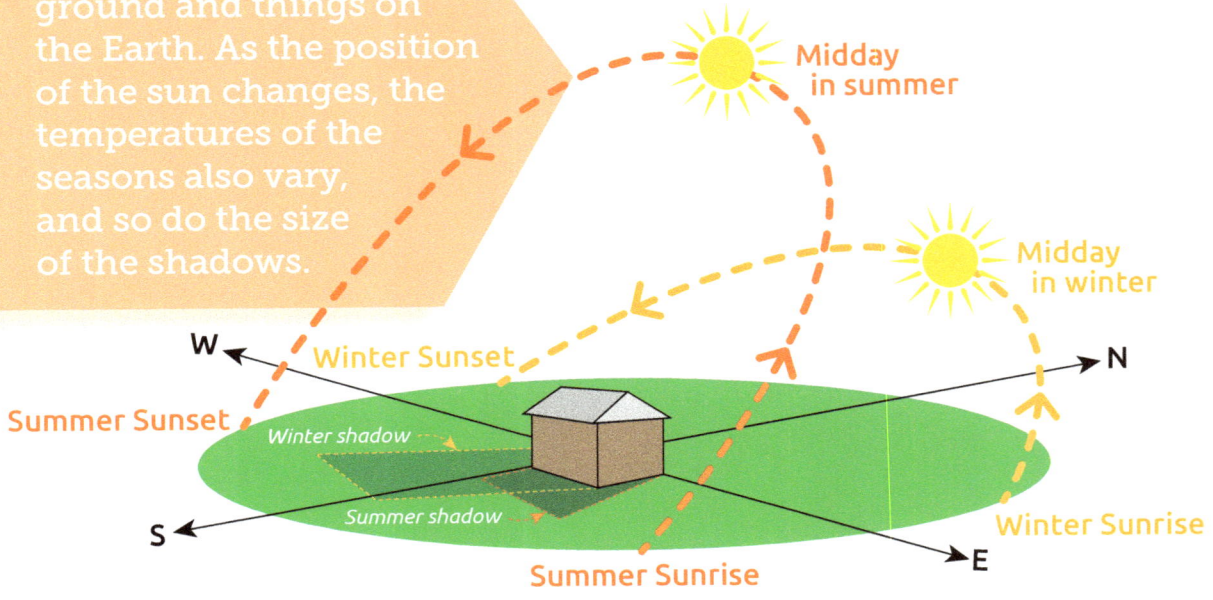

The sun's path in the southern hemisphere

Windows can be a major source of transfer of heat. Double or triple **glazing** is one solution. **Shading** is particularly important during summer to keep your house cool.

Q4c. Design shading for the glass and open areas of your tiny house. Your shading should block the sun in summer and allow the transfer of heat during winter.
Use the Internet to investigate the sun's angles for your location. Use a torch or light stand to test your design. Remember to test the house under different sun positions. Sketch your design idea and record your results on these pages.

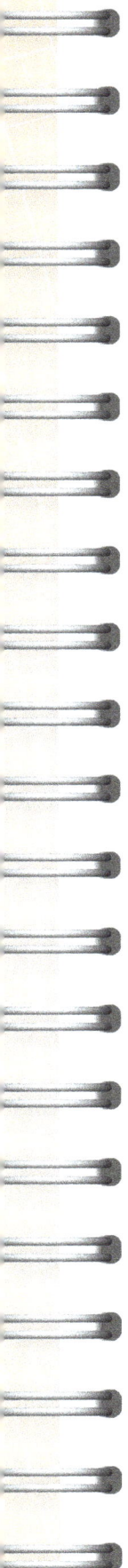

5. Improve

Present your final tiny-house idea to the class. Explain how your idea supports the United Nations Goal of Sustainable Cities and Communities.

Ask your classmates these questions:

- *What are some of the good points about my tiny-house idea?*
- *What are some of the weak points about my tiny-house idea?*

What did your classmates think?

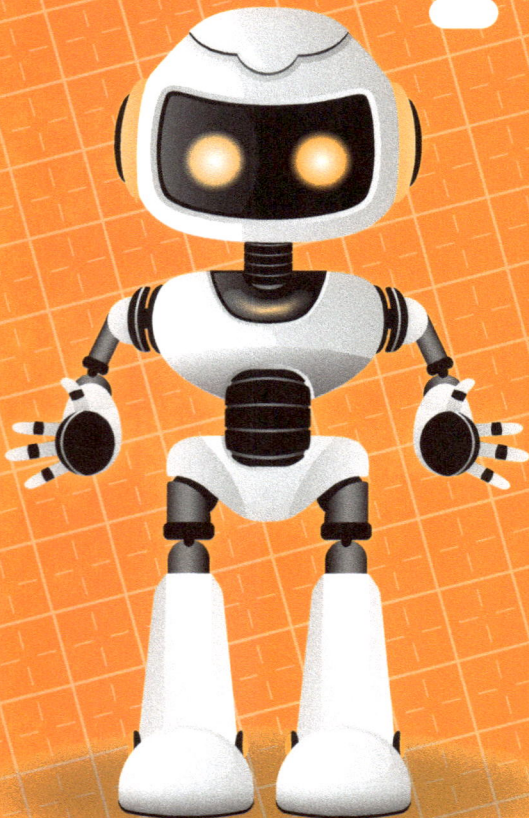

Q5a. Write your classmates' answers in the table.

Good points

Weak points

→ *Consider your classmates' feedback...*

Q5b. How could you improve your tiny-house idea?

_____
_____
_____
_____
_____
_____
_____
_____
_____

Now it's time to **Reflect...**

What have you learnt in this activity? *List three things.*
Hint: Look back at the key words.

1. _____

2. _____

3. _____

What would you like to know more about*?*

_____

_____

_____

_____

_____

_____

Curly Question

Estimate how much paint you would need if you had to paint the exterior of your tiny-house model. Model paints are sold in pots.

_____

_____

_____

_____

_____

_____

_____

_____

Challenge 6

An item made from waste

United Nations Sustainable Development Goal:

Climate Action

Action on **climate change** is one of the goals of the United Nations. What we do on Earth affects the climate. A huge amount of **urban waste** is dumped every day. This causes **pollution**, which can affect the world's climate. Waste is recycled, but an alternative to recycling is **upcycling**. Upcycling is taking urban waste and creating a new use or purpose for it.

Key words in this activity:

- climate change
- pollution
- upcycling
- urban waste
- brainstorm
- demolition

A project officer in the resource recoveries section at the local council wants to draw attention to the ever-growing issue of urban waste. Urban waste such as tyres, tubes, furniture, textiles, metals or packaging is being dumped daily. Recycling is a solution, but upcycling is a new goal. Your challenge is to upcycle a piece of urban waste. The best idea will be displayed in the library to promote upcycling.

1. Ask

Q1a. What does upcycling mean to you?

_____

_____

_____

Q1b. What does urban waste mean to you?

_____

_____

_____

_____

Q1C. Find two examples of upcycling in your community.

For each one, say how the urban waste was upcycled.

Information

i

Upcycling is better for the environment as materials and products are not broken down and pollution is reduced.

Example 1

An example of urban waste in my community is:

How was the urban waste upcycled?

Example 2

An example of urban waste in my community is:

How was the urban waste upcycled?

Search

r Read about wastes, upcycling and materials. You can gather information from books in your library or you can do a search on the Internet. Here are some websites and search terms to get you started.

1. Upcycling Ideas for Kids: https://bonzabins.com.au/upcycling-ideas-for-kids/ Search terms: Upcycling Ideas for Kids
2. Officeworks: https://www.officeworks.com.au/noteworthy/post/upcycle-it-fun-recycled-craft-ideas-for-kids   Search terms: Upcycling Ideas for Kids
3. SEE Library: https://seelibrary.org/ Look up the "Science and Technology" category. Search terms: seelibrary, science and technology

2. Imagine

*When you **brainstorm**, you put a group of people's ideas together.*

Q2a. By brainstorming, how many ideas can you come up with for the following urban wastes?

Used textiles: _____

_____

Discarded packaging: _____

_____

Old tyres: _____

_____

Q2b. Compare your group's answers with the answers of students in other groups. *Record a few interesting brainstorming ideas from other groups.*

_____

_____

_____

_____

_____

Q2c. What is your idea for upcycling urban waste?

The urban waste to be upcycled is:	*My idea is:*
_____	_____
_____	_____
_____	_____

3. Plan

Q3a. Use the grid to draw a rough sketch of your upcycling idea.

Q3b. Share your upcycling idea with your classmates. What do they think of your idea? Write their comments here.

→ *Consider your classmates' feedback...*

Q3c. Draw a final sketch of your upcycling idea. Remember to:
• label the materials
• add the **measurements**
• add any other information and details.

64

Q3d. In the table below, list the recyclable waste and quantities (No.) you will need for your upcycling idea.

Recyclable waste	No.

Safety

(S)

Safety is important to prevent accidents and injuries. Using tools and equipment can be dangerous. You should always wear safety gear to protect yourself when handling urban waste. The other term for safety gear is "Personal Protective Equipment" or "PPE".

Some examples of PPE: gumboots, protective gloves, and goggles.

→ *Procedure*

Q3e. How will you upcycle the urban waste? *Write down the steps you will follow.*

_____
_____
_____
_____
_____
_____
_____
_____
_____
_____
_____
_____

My STEM Workbook 2 – Understanding Science, Technology, Engineering and Mathematics through design-process activities   Vinesh Chandra and Basil Slynko   ISBN: 978-0-6484052-3-8

Q3f. What PPE should you wear when handling the six types of urban waste listed below?

1 **Concrete waste**

The PPE I should wear is:

_____

4 **Glass waste**

The PPE I should wear is:

_____

2 **Old tyres**

The PPE I should wear is:

_____

5 **Old batteries**

The PPE I should wear is:

_____

3 **Demolition materials**

The PPE I should wear is:

_____

6 **Food waste**

The PPE I should wear is:

_____

Artists and persons undertaking craft activities often upcycle urban waste. One only needs to look at the Internet or local markets to see the huge range of upcycled items for sale.

Q3g. Three upcycled items are shown below. Take some time to look at them. You may either 'Like' or 'Dislike' the item. *Record your evaluation below.*

I like/dislike this idea because:

_____

_____

_____

_____

I like/dislike this idea because:

_____

_____

_____

_____

I like/dislike this idea because:

_____

_____

_____

_____

There are two basic types of waste. They are:
- *hazardous*
- *non-hazardous.*

A number of hazardous household items are shown below.

a. Cleaning spirits; b. Lubricants; c. Insecticides; d. Paint, rust inhibitors and solvents; e. Silicone and sealants; f. Garden sprays.

Q3h. How should hazardous waste be disposed of? *Hint: Local council websites and manufacturer's product labels may have more information.*

_____
_____
_____
_____
_____
_____
_____
_____

4. Create
Upcycling an urban-waste item

Remember to ask questions and seek feedback from your teacher while you create your upcycled urban-waste model.

You need to:
- gather the urban waste
- gather the tools and equipment you need
- follow your procedure to upcycle the urban waste
- use PPE and follow safety rules.

Q4a. Upcycling an urban-waste item may not go as planned. For each step that caused problems, explain how you dealt with it. List the step(s) and your course of action.

_____
_____
_____
_____
_____
_____
_____
_____
_____
_____
_____
_____
_____
_____
_____
_____
_____
_____

Challenge 6. An item made from waste

Class Survey

Q4b. Ask 10 students in your class what they think of your upcycled item. Place a tick (✓) in the column of their choice, and tally the results at the bottom.

Student's name	Very good	Good	OK	Needs work
Total:				

Q4c. Use the totals from your survey above to complete a column graph. →

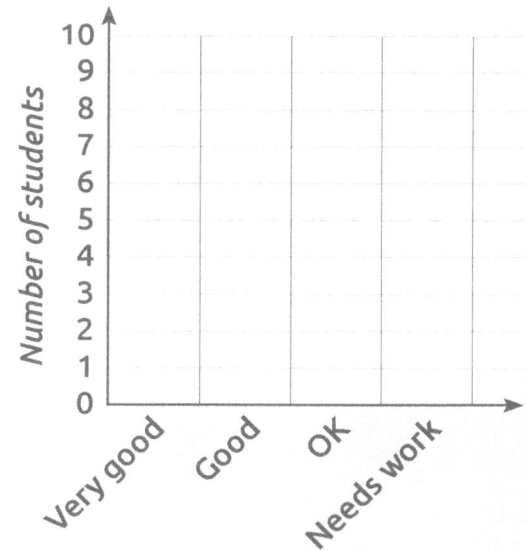

Number of students

10
9
8
7
6
5
4
3
2
1
0

Very good   Good   OK   Needs work

→ *Interpret*

Q4d. What did the survey tell you?

_____

_____

_____

5. Improve

Present your final upcycling idea to the class. Explain how your idea supports the United Nations Goal of Climate Action.

Ask your classmates these questions:

- *What are some of the good points about my upcycling idea?*
- *What are some of the weak points about my upcycling idea?*

What did your classmates think?

Q5a. Write your classmates' answers in the table.

Good points
Weak points

→ *Consider your classmates' feedback...*

Q5b. How could you improve your upcycling idea?

_____

_____

_____

_____

_____

_____

_____

_____

_____

Now it's time to Reflect...

What have you learnt in this activity? *List three things.*
Hint: Look back at the key words.

1. _____

2. _____

3. _____

What would you like to know more about?

_____

_____

_____

_____

_____

_____

Curly Question

What would have happened to the urban waste if you had not upcycled it? Think about its 'life cycle' until it fully breaks down.

_____

_____

_____

_____

_____

_____

_____

_____

Challenge 7

A digital book on plastics in waterways

United Nations Sustainable Development Goal:

Life below Water

For the United Nations, caring for our seas, oceans and other waterways is a priority. We need the plants and animals that live in these waters for our survival. Over the years there has been a lot of damage done to waterways that needs to be repaired. There is a need to effectively manage and protect the **habitat** of all living creatures that live below water.

Key words in this activity:

- ocean • habitat
- decomposer
- producer • plastic
- consumer
- organism

A publisher wants you to create a digital book about organisms that live below water. In your book you should write about marine habitats and food chains and how they connect consumers, producers and decomposers that live under water. You also need to say how we can protect and take care of organisms that live below water. You can share your book with your class and other children in the world through an online library.

Note: You can use pictures and drawings in your book. Do not include any photos of people. Your book needs to be between 6 and 10 pages.

1. Ask

Q1a. **What is an organism?**

_____

_____

_____

_____

_____

_____

Q1b. What does the term 'habitat' mean?

_____

_____

Q1c. In the table below, explain what these terms mean and give some examples of each.

Category	What does the term mean?	Give some examples
Producer		
Consumer		
Decomposer		

Read about life under water.

You can gather information from books in your library or you can do a search on the Internet. Here are some websites and search terms to get you started.

1. Primary Connections: https://www.primaryconnections.org.au/ themes/custom/connections/assets/SBR/data/Bio/sub/foodchain/foodchain.htm Search terms: primary connections, food chain

2. National Geographic Kids: https://kids.nationalgeographic.com/nature/habitats/ article/ocean   Search terms: national geographic kids, **ocean** habitats

3. SEE Library: https://seelibrary.org/ Look up the "Science and Technology" category. Search terms: seelibrary, science and technology

2. Imagine

Q2a. Continue your research and find information that will help you answer the questions in the table below. Jot down some dot points. These are the points that you can use in your book.

---

1. What is a marine habitat?

2. What are some examples of marine organisms?

3. What are some examples of marine food chains?

4. What are some dangers that marine organisms face?

5. Why do marine organisms need clear waterways?

6. How can marine waterways be damaged?

7. How can marine life be managed and protected?

Q2b. This picture shows some marine organisms. *Use the Word Bank to identify the organisms, then label each organism as either a* **decomposer** (D), **producer** (P), *or* **consumer** (C). *Draw arrows over the picture to show a food chain that connects a decomposer, a producer and a consumer.*

Word Bank

algae   bird   crab   fish
prawn   weeds   worm

	Organism	D/P/C
1		
2		
3		
4		
5		
6		
7		

You will use an application called **Book Creator** *to write your book. Your teacher will explain how this application works.*

Search "The SEE Digital Library" (https://seelibrary.org) to view some examples of digital books.

Information

i

Q2c. Do you have any questions about Book Creator?
List your questions here and ask your teacher.

_____

_____

_____

_____

Challenge 7. A digital book on plastics in waterways

Page 1

You will need a storyboard.
A storyboard is a plan for a story.
It shows what will go on each page.

Information

i

Page 2

3. Plan

Q3a. Write your story in this storyboard using a pencil so you can make changes as you think of new ideas.

Front cover

Page 3

Page 4

Page 7

Page 5

Page 8

Page 6

Back cover

→ *Procedure*

Q3b. Write a procedure for creating your digital book.

_____
_____
_____
_____
_____
_____
_____
_____
_____
_____
_____
_____
_____
_____
_____
_____
_____
_____
_____
_____
_____
_____

Tools and Equipment

Q3c. What equipment will you use to create your book?

_____
_____
_____
_____
_____
_____

Internet safety is very important. Seek help from your teacher or your parents when using the Internet. Get their advice on what websites to view. Do not click links that you are not sure of. Do not share your details online or connect with people you do not know. When in doubt talk to your teacher or parents.

Q3d. What safety rules do you need to follow when using the Internet? *List the three rules that you think are the most important.*

Rule 1: _____
_____
_____

Rule 2: _____
_____
_____

Rule 3: _____
_____
_____

4. Create
Making a digital book

Remember to ask questions and seek feedback from your teacher while you create your digital book.
You need to:

- log in to the Internet (follow your teacher's instructions)
- follow your procedure
- use your storyboard to create your book
- follow your Internet safety rules at all times.

Q4a. Making your digital book may not go as planned. For each step that caused problems, explain how you dealt with it. List the step(s) and your course of action.

_____

_____

_____

_____

_____

_____

_____

_____

_____

_____

_____

_____

_____

Q4b. Read your book to an adult or an older sibling or friend. Ask them these questions and write your answers as dot points below.

What did you think of my book?

What did you learn from my book?

How can I improve my book?

5. Improve

Present your digital book to the class. Explain how your book supports the United Nations Goal of Life below Water.

Ask your classmates these questions:

- *What are some of the good points about my digital book?*
- *What are some of the weak points about my digital book?*

What did your classmates think?

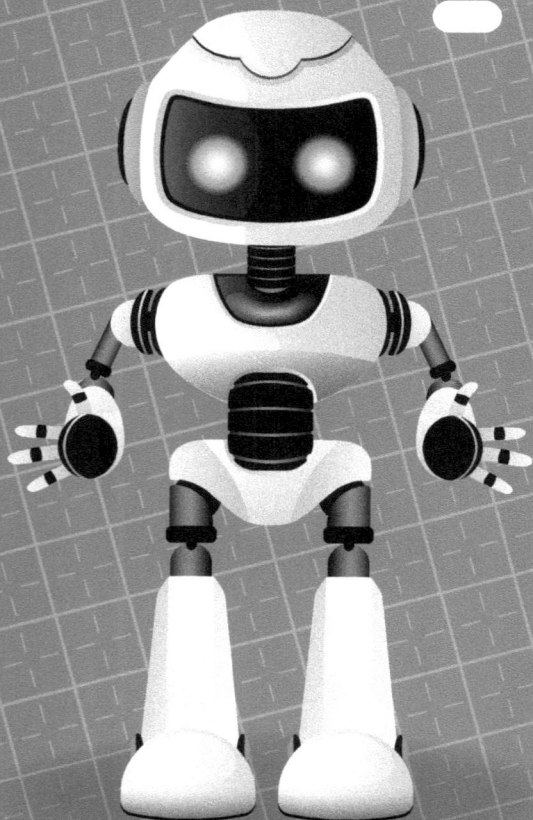

Q5a. Write your classmates' answers in the table.

Good points

Weak points

→ *Consider your classmates' feedback...*

Q5b. How could you improve your digital book?

_____

_____

_____

_____

_____

_____

_____

_____

Challenge 7. A digital book on plastics in waterways

Now it's time to Reflect...

What have you learnt in this activity? *List three things.*
Hint: Look back at the key words.

1. _____

2. _____

3. _____

What would you like to know more about?

_____

_____

_____

_____

_____

_____

Curly Question

Plastics in water are very harmful to marine life. Researchers believe that the amount of **plastic** we produce in the world doubles every 20 years. How can this harm our marine life?

_____

_____

_____

_____

_____

_____

_____

_____

Challenge 8

A shelter for chickens

United Nations Sustainable Development Goal:

Life on Land

Caring for the life of all living things on Earth is an important goal of the United Nations. All living things need shelter. Shelters protect living things from the weather and predators, and they are a place for them to raise their young.

Key words in this activity:

- living things
- farm-to-plate
- design
- form
- flowchart
- life cycle
- dimensions

There is talk at home about sustainable living. The goals are to reduce waste – food and plant – and use a **farm-to-plate** approach at home. One solution is to get chickens and breed them. Your task is to plan and build a model of a shelter for keeping and breeding chickens. You will decide how many chickens and chicks your shelter will hold. The shelter could also be movable.

1. Ask

Q1a. What sorts of things do you need to think about when designing a shelter for chickens?

_____

_____

_____

_____

_____

_____

_____

_____

Q1b. What should be inside the chicken shelter? *List the features and their main purpose below.* One example is listed.

Feature	Main purpose
Feeder	To hold the grain

Q1c. What are the dimensions of a chicken? *Use the Internet for information, then record the measurements here.*

Length: _____ Width: _____ Height: _____
(beak to tail)   *(wing to wing)*   *(comb to feet)*

r   **Read about animal shelters and materials.**
You can gather information from books in your library or you can do a search on the Internet. Here are some websites and search terms to get you started.

1. YouTube: https://www.youtube.com/watch?v=JpFgxo0K_TE
   Title of video: P.3, Science, Life Cycle of a Chicken
2. Bunnings: https://www.workshop.bunnings.com.au/t5/Outdoor/DIY-chicken-coop/td-p/111727   Search terms: DIY chicken coop
3. SEE Library: https://seelibrary.org/ Look up the "Science and Technology" category. Search terms: seelibrary, science and technology

2. Imagine

All animals have a life cycle.

Q2a. *Draw a diagram to show the **life cycle** of a chicken.*

i *Information*

Materials can be sorted into two groups: natural and man-made. *See page 94 for more information on materials.*

Q2b. What natural and man-made materials could you use to build a chicken shelter? *List two materials for each part of the shelter.* Why did you choose these materials?

Parts of the chicken shelter	Why did you choose these materials?
Roof: 1. 2.	
Walls: 1. 2.	
Floor: 1. 2.	

Q2c. What **form** could you use for your chicken shelter? *Sketch three examples and write the name of the form underneath.*

The form is:

_____

The form is:

_____

The form is:

_____

Q2d. You need to consider the welfare of the chickens and chicks. Think about the overall size of your chicken and chicks' shelter. *Consider how many chickens you want it to house.*

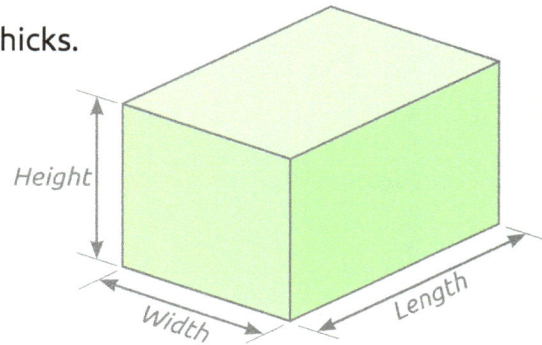

Number of chickens: _____   Number of chicks: _____

What would be the dimensions of your chicken shelter?

Length: _____ cm   Width: _____ cm   Height: _____ cm

Q2e. Consider the shape of the roof. *Sketch a few designs on the grid.*

Q2f. *Complete the table below.* List the materials you will need to build your chicken shelter, and record the main properties and weaknesses of each material.

Suitable material	Main properties	Weakness(es)
Roof: 1. 2.		
Walls: 1. 2.		
Floor: 1. 2.		

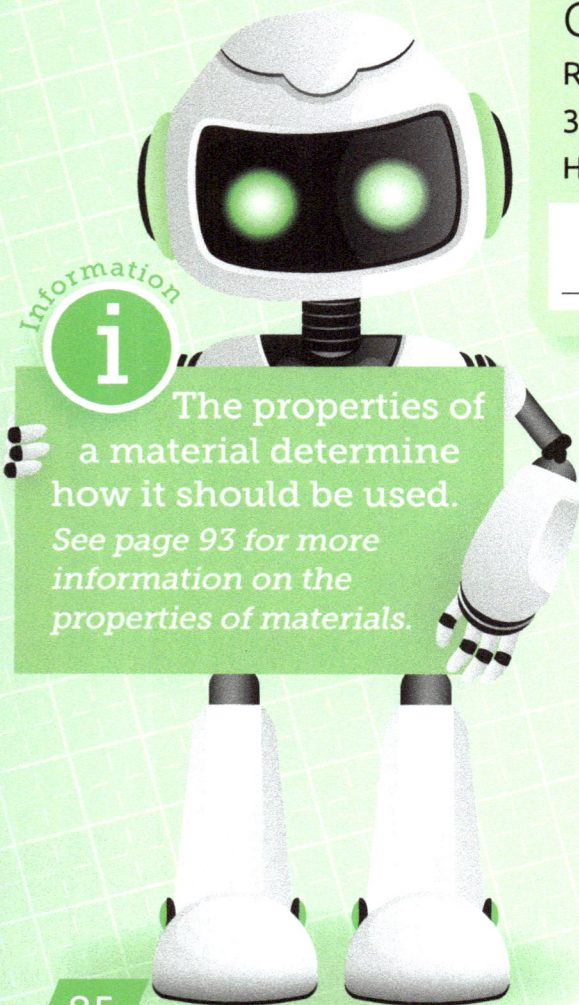

Q2g. According to the RSPCA, a hen will lay about 300 eggs in the first year. How much is 300 in dozens?

_____

Q2h. If you could sell the eggs for $6.50 a dozen, how much income would it generate?

$ _____

Information

ⓘ The properties of a material determine how it should be used. *See page 93 for more information on the properties of materials.*

Q2i. Keeping chickens in a coop costs money. What would be some of the expenses?

_____

_____

_____

_____

_____

3. Plan

Q3a. Use the grid to sketch your initial idea for a shelter for chickens.

Q3b. Share your chicken shelter idea with your classmates. What do they think of your idea? Write their comments here.

_____

_____

_____

_____

_____

_____

_____

_____

_____

_____

_____

_____

_____

_____

→ *Consider your classmates' feedback...*

Q3c. Can your shelter idea be improved on? Some things to consider are the welfare of the chickens, shape, size, disposal of waste and access. *Draw a final sketch below of your chicken shelter idea.* Remember to:
• label the materials
• add the measurements in centimetres
• add any other information and details.

Q3d. In the table below, list the materials and quantities (No.) you will need for your chicken shelter.

Material	No.	Material	No.

Word Bank

bedding   features   finish   frame

materials   roof   shelter   site

Q3e. How could you make the chicken and chick shelter? *Fill in the blanks in the flowchart using the words from the Word Bank.*

Step	Process
1	Select a _____
2	Mark out the _____
3	Get the _____
4	Build the _____
5	Install the _____
6	Add the _____
7	Apply the _____
8	Add the _____

Safety

S Safety is important to prevent accidents and injuries.
Using tools and equipment can be dangerous when building things.
You should always wear safety gear to protect yourself.
The other term for safety gear is "Personal Protective Equipment" or "PPE".

Q3f. Three processes to make a chicken and chick shelter are listed below. *List the health hazard(s) for each process and the PPE you should wear to prevent each health hazard.*

1 Carrying sheet metal

Health hazard(s): _____

_____

PPE I should wear: _____

_____

2 Cutting plywood

Health hazard(s): _____

_____

PPE I should wear: _____

_____

3 Fixing chicken-wire mesh

Health hazard(s): _____

_____

PPE I should wear: _____

_____

4. Create
Making a model of a chicken shelter

Remember to ask questions and seek feedback from your teacher while you create your chicken shelter model.

You need to:

- gather all the materials, tools and equipment
- follow your flowchart steps as a guide to make your model of the chicken shelter
- use PPE and follow safety rules.

Q4a. Making your chicken shelter may not go as planned. For each step that caused problems, explain how you dealt with it. List the step(s) and your course of action.

_____

_____

_____

_____

_____

_____

_____

_____

_____

_____

_____

_____

_____

_____

Test

t

Q4b. Use your model to test the level of comfort. Work with classmates to come up with an idea(s) for how to test the level of comfort. *Record your comments on these pages.*

Challenge 8. A shelter for chickens

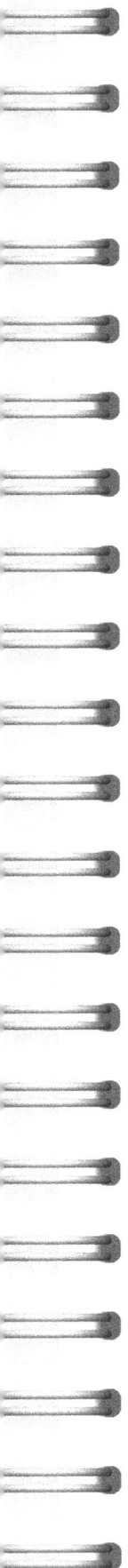

5. Improve

Present your final chicken shelter idea to the class. Explain how your idea supports the United Nations Goal of Life on Land.

Ask your classmates these questions:

- *What are some of the good points about my chicken shelter idea?*
- *What are some of the weak points about my chicken shelter idea?*

What did your classmates think?

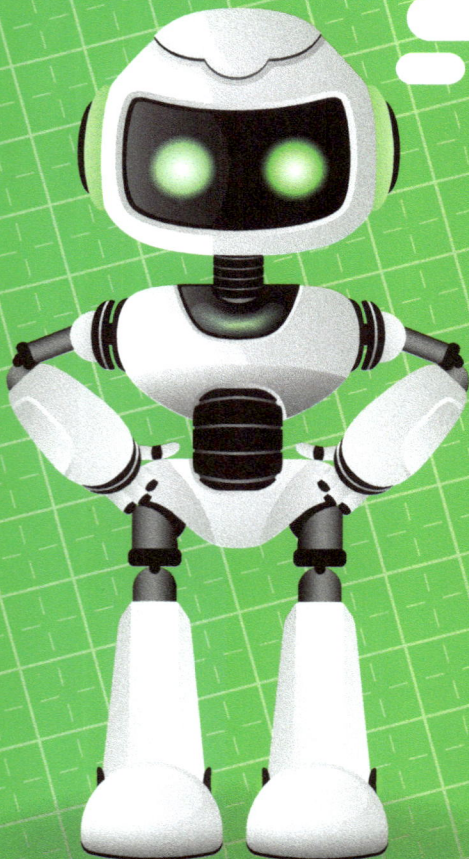

Q5a. Write your classmates' answers in the table.

Good points

Weak points

→ *Consider your classmates' feedback...*

Q5b. How could you improve your chicken shelter idea?

_____

_____

_____

_____

_____

_____

_____

Now it's time to Reflect...

What have you learnt in this activity? *List three things.*
Hint: Look back at the key words.

1. _____

2. _____

3. _____

What would you like to know more about?

_____
_____
_____
_____
_____
_____

Curly Question

How can you **design** your chicken shelter idea as a flat-pack so it is easy to dismantle and transport? *Record your idea below. Use a grid (pages 95-98) to sketch your solution.*

_____
_____
_____
_____
_____
_____
_____
_____
_____

Some properties of materials

Different materials are used for different things. Have you ever wondered why? It is to do with the properties of materials.

Each material has its own set of properties. These properties determine how a material should be used. For example, wood is strong when squeezed (compression) and steel is strong when stretched (tension).

Hardness
How well does it resist denting or scratching?

Strength
How easy or difficult is it to break?

Toughness
How much force can it stand without breaking?

Density
How heavy or light is it for its size?

Elasticity
How well does it return to its original shape and size after being subjected to external loads and forces?

Flammability
How easily does it burn or smoulder when exposed to heat or fire?

Conductivity
How well does it conduct or insulate against heat, electricity or sound?

Durability
How well does it resist environmental stresses such as weather or insects, or does it tear easily or wear out (e.g. fabric)?

93

Source: Design and Technology in Today's World: A First Look, Years 3–6; Basil Slynko.

Some samples of natural and man-made materials

Materials can be sorted into four basic groups: natural, processed, synthetic and composite.

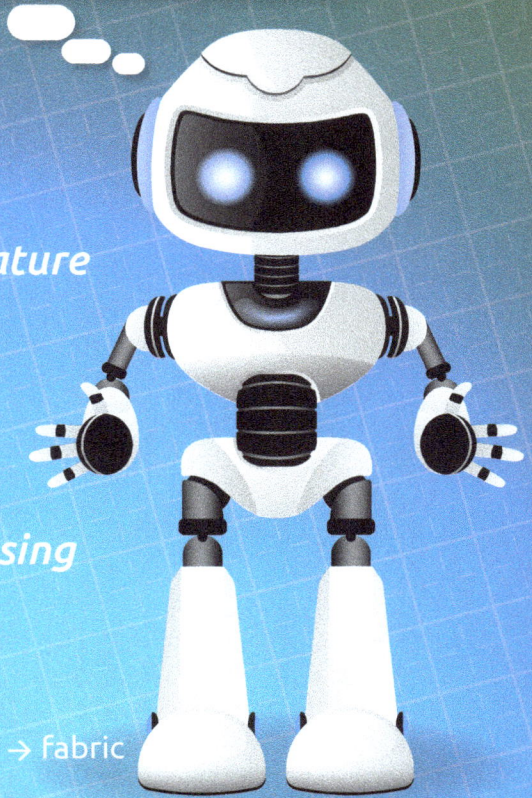

Natural – found in nature
a. Wood
b. Stone
c. Mud
d. Cork
e. Straw

Processed – usually using natural materials
f. Ore → steel
g. Bauxite → aluminium
h. Clay → ceramics
i. Plant or animal fibres → fabric

Synthetic – made from chemicals
j. Acrylic sheeting
k. Polystyrene
l. Polyester resin
m. Epoxy resin

Composite – combining two or more materials
n. Cob = clay + sand + straw
o. Gore-Tex® = fabric + plastic film + thin layer of foam
p. Concrete = sand + stone + cement + water
q. Brass = copper + zinc

94

Notes

Notes

Notes

97

Notes and grid

Notes

www.ingramcontent.com/pod-product-compliance
Lightning Source LLC
Chambersburg PA
CBHW041101050426
42334CB00063B/3283